ACC. No.	CLASS. No.
245528	576.8 STE
	CHECKED.

KU-006-421

DAWKINS vs. GOULD

Dawkins Vs Gould: Survival of the Fittest

245528

Non-Fiction   576.8 STE

Royal Grammar School Guildford

KIM STERELNY
DAWKINS
vs. GOULD
Survival of the Fittest

Icon Books

This edition published in the UK in 2007 by
Icon Books Ltd, The Old Dairy,
Brook Road, Thriplow,
Cambridge SG8 7RG
email: info@iconbooks.co.uk
www.iconbooks.co.uk

Originally published in the UK in 2001 by Icon Books Ltd

Sold in the UK, Europe, South Africa and Asia
by Faber & Faber Ltd, 3 Queen Square,
London WC1N 3AU
or their agents

Distributed in the UK, Europe, South Africa and Asia
by TBS Ltd, TBS Distribution Centre, Colchester Road
Frating Green, Colchester CO7 7DW

This edition published in Australia in 2007
by Allen & Unwin Pty Ltd,
PO Box 8500, 83 Alexander Street,
Crows Nest, NSW 2065

This edition published in the USA
in 2007 by Totem Books
Inquiries to Icon Books Ltd.,
The Old Dairy, Brook Road, Thriplow,
Cambridge SG8 7RG, UK

Distributed to the trade in the USA
by National Book Network Inc.
4501 Forbes Boulevard, Suite 200
Lanham, Maryland 20706

Distributed in Canada by
Penguin Books Canada,
90 Eglinton Avenue East, Suite 700,
Toronto, Ontario M4P 2YE

ISBN-10: 1-84046-780-0
ISBN-13: 978-1840467-80-2

Text copyright © 2007 Kim Sterelny

The authors have asserted their moral rights.

No part of this book may be reproduced in any form, or by any
means, without prior permission in writing from the publisher.

Typesetting by Hands Fotoset

Printed in the UK by CPI Bookmarque, Croydon, CR0 4TD

Contents

For Peter: friend and colleague

Acknowledgements to the first edition

Thanks to Mike Ridge, Matteo Mameli, Darryl Jones, Jochen Brocks, Dan Dennett and Jon Turney for their thoughts on a draft of this book. Thanks also to the Philosophy Program, Research School of the Social Sciences, Australian National University and the Department of Philosophy, Victoria University of Wellington, whose support made it possible for me to write it.

Acknowledgements to the second edition

I am indebted to Peter Godfrey-Smith, Brett Calcott, Ben Jeffares, Jochen Brocks, Lindell Bromham and the other members of Tempo and Mode for encouraging me to write this second edition, and for their advice as I did so. I am also indebted to Mike Hannah, Geoff Chambers, John Endler, Russell Gray, David Penny and (especially) Chris Clowes for their constructive responses to the first edition, responses which guided me as I wrote the second version. I have institutional debts as well: the philosophy programmes at the Australian National University and the Victoria University of Wellington, my two academic homes, continue to be friendly, civilised and supportive environments. Their support makes my work possible.

About the author

Kim Sterelny is the author of several books, including *Sex and Death* (University of Chicago Press, 1999) and *Thought in a Hostile World* (Blackwell, 2003), which won the Lakatos Prize in 2004. He works in the philosophy departments at Victoria University in Wellington, New Zealand and the Australian National University, Canberra. In addition to philosophy, Kim's interests include eating curries, drinking red wine, bushwalking and birdwatching.

Part I
Battle Joined

1 A Clash of Perspectives

Science in general, and biology in particular, has seen its fair share of punch-ups. In the 1930s and 1940s, Britain's two greatest biologists, J.B.S. Haldane and R.A. Fisher, feuded so vigorously that their students (John Maynard Smith tells me) were hardly allowed to talk to one another. But their behaviour was civilised compared to the notorious feuds in biological systematics between cladists – notorious for wielding unintelligible terminology and vituperation in equal measure – and their opponents. Mostly these fights are kept more or less in-house, often because the issues are of interest only to the participants. Almost no one except systematicists are interested in the principles by which we tell that *Drosophila subobscura* is a valid species. But sometimes these disputes leak out into the open. Richard Dawkins and Stephen Jay Gould have different views on evolution, and they and their allies have engaged in a public and polemical exchange, though on one side, at least, this debate must now be carried on through proxies. Gould died in 2002, shortly after his enormous *The Structure of Evolutionary Theory* appeared.

At first glance, the heat of this exchange is puzzling. For Dawkins and Gould agree on much that matters. They agree that all life, including human life, has evolved over the last 4 billion years from one or a few ancestors, and that those first living things probably resembled living bacteria in their most crucial respects. They agree that this process has been wholly natural; no divine hand, no spooky interloper, has nudged the process one way or another. They agree that chance has played a crucial role in determining the cast of life's drama. In particular, there is nothing inevitable about the appearance of humans, or of anything like humans: the great machine of evolution has no aim or purpose. But they also agree that evolution, and evolutionary change, is not just a lottery. For natural selection matters too. Within any population of life forms, there will be variation. And some of those variants will be a touch better suited to the prevailing conditions than others. So they will have a better chance of transmitting their distinctive character to descendants. Natural selection was one of the great discoveries in Darwin's *Origin of Species* (1859). If a population of organisms varies one from another; if the members of that population differ in fitness, so one is more likely than another to contribute her descendants to the next generation; if those differences tend to be heritable, so the fitter organism's offspring share her special characteristics, then the population will evolve by natural selection. Australia is renowned for its poisonous snakes and of these the taipan is the most famously venomous. Let's consider the mechanism through which it became so impressively lethal. If a population of taipans differ in the toxicity of their venom; if the more venomous snakes survive and reproduce better than less venomous ones, then taipans will, over time, evolve more toxic venom. Gould and Dawkins agree that complex capacities like human vision, bat echolocation or a snake's ability to poison its prey evolve by natural selection. And they agree that in human terms, natural selection works slowly, over many generations. Bacteria and other single-celled organisms whip through their generations at speed, and that is why drug resistance outpaces new drugs. But for larger, more slowly reproducing organisms, significant changes take thousands of years to build.

Adaptive change depends on cumulative selection. Each generation is only slightly different from the one that precedes it. It was once received wisdom that the pace of evolutionary change had to be very slow, for the co-adapted construction of organisms constrains the extent to which one organ can change without correlated changes in others. Change must be gradual, for the parts of an organism are delicately and precisely adjusted to one another, so almost all large, random changes are disasters. Adding a horn to a horse's head might seem to provide it with a useful defensive weapon, but without compensating changes to its skull and neck (to bear the extra weight) it would be not only useless but detrimental. It is now widely agreed that the effect of this constraint has been exaggerated. This line of thought overlooks the importance of phenotypic plasticity. The developmental resources an organism inherits do not determine its final form. It makes a difference whether people are born on the coast or in the high mountains; near the equator or the poles. As we grow, our physiology adapts us to our specific environment; a child growing up in a hot climate has more sweat glands than one born in a cold climate. Well-fed and well-exercised children develop greater muscle mass. And they also develop bones and circulatory systems to support those muscles. In development, one organ system responds to signals from others: they adjust to one another. We are typical animals in this respect. So a horse sporting a genetically caused change in its armoury might not have to wait for new genes to build stronger muscles and bones. (Marc Kirschner's and John Gerhart's *The Plausibility of Life* is a splendid account of phenotypic plasticity and its evolutionary importance.) Even so, the ability of an organism's developmental systems to cope with the unexpected is limited. Perhaps, very occasionally, a major evolutionary change appears in a single generation, as the result of one big mutation. But such events will be rare. So large single-step changes, Gould and Dawkins agree, must be very rare. The normal history of an adaptive invention is a long series of small changes, not a short series of large changes.

Yet despite these agreements on fundamentals, Dawkins and Gould have clashed heatedly on the nature of evolution. In two notorious articles in *New York Review of Books*, Gould scathingly reviewed *Darwin's Dangerous Idea*, a work of Dawkins' intellectual ally Daniel Dennett. In 1997, there was a better-tempered but no more complimentary exchange in *Evolution*, as they traded reviews of each other's most recent creation. Dawkins and Gould are representatives of different intellectual and national traditions in evolutionary biology. Dawkins' doctoral supervisor was Niko Tinbergen, one of the co-founders of ethology. Ethology aims to understand the adaptive significance of particular behavioural patterns. So Dawkins' background sensitised him to the problem of adaptation; of how adaptive behaviours evolve in a lineage and develop in an individual. Gould, in contrast, is a palaeontologist. His mentor was the brilliant but notoriously irascible George Gaylord Simpson. The match, if it exists, between an animal's capacities and the demands of its environment is less obvious with fossils than with live animals. A fossil gives you less information on the animal and its environment. Palaeobiologists are often historians of life: they are trained to see the similarities among their fossilised organisms that help them understand patterns of descent. So it is tempting to suppose that the passion of these exchanges reflects nothing deeper than competition for the same patch of limelight, magnified by different historical and disciplinary perspectives. I think that suspicion would be misplaced, and it's my aim in this book to explain why. Despite real and important points of agreement, their clash is of two very different perspectives on evolutionary biology. Those competing perspectives persist, despite the death of one spokesperson, so it is worth revisiting these issues in this second edition in the light of recent developments in evolutionary biology.

For Richard Dawkins, the fit between organisms and environment – adaptedness, or good design – is the central problem evolutionary biology must explain. He is most struck by the problem Darwin solved in his *Origin*: in a world without a divine engineer, how can complex adaptive structures come into existence? In his view, natural selection is the only possible answer to this question. Natural selection is the only natural mechanism that can produce complex, co-adapted structures, for such structures are vastly improbable. Hence natural selection plays a uniquely important role in evolutionary explanation. Moreover, and most famously, Dawkins argues that the fundamental history of evolution is the history of gene lineages. The molecular biology of genes – the chemical details of their action, interaction and reproduction – is alarmingly complex. But fortunately Dawkins does not allow himself to be bogged down in these details, and we can follow his lead. He argues that the critical agents in life's drama must persist over long periods *precisely because* the invention of adaptation requires a long series of small changes. Hence the targets of selection are lineages that persist over many generations, lineages in which each link is identical or almost identical to its predecessor. Gene lineages and only gene lineages satisfy this condition. Genes are replicated: there are mechanisms that copied some of my genes into my daughter's genome; and those same mechanisms are capable of copying those same genes generation by generation. So genes form lineages of identical copies. These lineages can have very deep histories. You have genes that you share with yeasts and other single-celled organisms; organisms that have been evolving separately for billions of years. Perhaps with the exception of those organisms that clone themselves, organisms do not form lineages of identical copies. Reproduction is not copying. My daughter is not a copy of me. An organism is unique and ephemeral. So it disappears at the end of its life. But an organism's genes need not disappear. If that organism, or a relative carrying a similar complement of genes, reproduced, then copies of the organism's genes will persist. They may do so for many generations.

Moreover, the chance that a gene will be copied is not independent of the character of that gene. It is true that some genes are *silent*, and just seem to be hitching a ride. But often genes influence their own replication prospects. They do so most overtly by their influence on the characteristics (the *phenotype*) of the organism that bears them. So genes influence their own prospects of being copied. Dawkins conceives of the fundamental struggle of evolution as the struggle of genes in lineages to replicate. Moreover, success for one gene lineage can mean failure for another. Dawkins' opponents often portray him as a crazed reductionist, thinking that only genes matter in evolution. That is not his view. Organisms are important, but primarily as weapons in the struggle between gene lineages. Gene lineages usually compete with other gene lineages by forming alliances. Rival alliances build rival organisms. Successful organisms replicate the genes in the alliance that builds them. Thus macaw-making genes which build macaws suitable to the bird's circumstances become more common over time. The conflict between two macaws for a safe hollow in which to nest influences evolution by determining which lineages of macaw-making genes will be represented in the next generation, and in what numbers. The ecological struggle between organisms to survive and reproduce is translated into differential copying rates for the genes that build those organisms.

In short, for Dawkins, the history of life is a history of a war between gene lineages. The beautiful biological mechanisms that we see on so many natural history documentaries are the visible products of that war. They are its weapons. For rival gene alliances are engaged in a perpetual arms race. In human arms races, weapons improve over time. So too will biological weapons, though that improvement has from time to time been disrupted by catastrophic and unforeseeable changes to the battlefield: episodes of mass extinction when many species disappear. These changes may be caused by the geology of the earth itself, as continents divide, mountains erupt, seas and ice fields advance or retreat. And they may be caused by forces external to the earth: by impact or by changes in the sun's behaviour. But between these episodes, selection is omnipresent and effective, sifting gene teams, building adaptive improvements in the organisms that are their *vehicles*, as Dawkins puts it.

Gould sees the living world very differently. Life today is fabulously diverse. But many forms of life that used to dominate their environments are no longer with us. Gould is a palaeontologist, and so much of his professional life concerns extinction: from the spectacular extinction of the dinosaurs, pterosaurs and huge marine reptiles, to the less obtrusive, and yet in Gould's eyes more fundamental, extinctions of weird marine invertebrates 500 or more million years ago. The first multi-celled animals in the fossil record lived from about 570 million years ago to the beginning of the Cambrian Period (now dated to 543 million years ago), after which, they disappeared. The fossils of this so-called 'Ediacaran fauna' consist of the remains of frond- and disc-shaped organisms, and interpretations of these fossils vary widely. Some think they are more like lichen than animals. After the Ediacaran disappearance, in the Cambrian Period, for the first time we have a clear fossil record of the main modern lineages. By then, arthropods (insects, crabs and their kin) had evolved. So had bivalves (oysters, clams and the like) and gastropods (snails and their relatives). Jellyfish and sponges were around too, though they may have appeared a little earlier than the others. A horde of the different kinds of worms appeared. So too did the first chordates; our group. But at the same time, many other lineages came into existence, only to go extinct again. The Cambrian ended 490 million years ago, and by that time, the lights had gone out for many strange-seeming animal lineages.

Dawkins is impressed by the power of selection to build adaptations. Gould is equally struck by conservative aspects of the history of life. In their most fundamental ways, animal lineages do not seem to change over enormous stretches of time. There are hundreds of thousands, perhaps millions, of species of beetle. Every single one is built on the same basic plan. They vary in size, colour, sexual ornamentation and much else. But they are all recognisably beetles. The same is true of the other great lineages of animal life. The main division of the kingdom of animals is into *phyla*. There are 30-odd: the exact number is in dispute. Some are not known as fossils at all. James Valentine, in his splendid *On the Origin of Phyla* (p. 186), lists twelve phyla (all of small, soft-bodied animals) with no fossil record. But all of those with decent fossil records appeared early. Moreover, as we shall see in Chapter 10, there is indirect evidence that the phyla without fossil records are also old. All this leads Gould to the view that the main ways of building an animal were all invented at roughly the same time, and no new ones – no new fundamental body organisations – have been invented since. Evolution has certainly not ground to a halt when it comes to inventing new adaptations. But if Gould is right, it does seem to have ground to a halt in inventing new phyla of animals. Gould sees this as the most striking fact evolutionary theory must explain.

Moreover, Gould has a different conception of the mechanism of evolution. His views differ from those of Dawkins in three important ways. Chance is more important. Genes are less important. The changing fluctuations of gene lineages record success and failure, but do not cause success and failure. And selection is relatively less powerful. For it needs a good supply of variation: selection can act only to magnify and sculpt variations that are found in the population, and Gould thinks that the supply of variants is limited. Let's begin with chance. It is important because mass extinction is important. In times of mass extinction, many species disappear, and the biological landscape is altered in fundamental ways, often creating unpredictable opportunities for lucky survivors. Survivors are lucky, because surviving, in Gould's view, depends more on luck than on fitness. So in explaining evolutionary history, Gould places less weight on selection than does Dawkins. Moreover, he has a different view of the way selection works. He is very sceptical of gene selection, for he doubts that particular genes usually have a consistent enough effect on the fitness of their bearers for Dawkins' story to make sense. The effect of a particular gene on a body depends on the other genes in that body and on many features of the environment in which the organism develops. So Gould thinks that when selection acts, it typically acts on individual organisms.

These differences within evolutionary theory are exacerbated by different assessments of science itself. As *Unweaving the Rainbow* and *A Devil's Chaplain* show, Dawkins is a wholehearted son of the Enlightenment. We should embrace the scientific description of ourselves and our world, for it is true (or the nearest approach to truth of which we are capable), beautiful and complete. It leaves nothing out. Gould does not think that science is complete. The humanities, history and even religion offer insight into the realm of value – of how we ought to live – independent of any possible scientific discovery. And while Gould has never embraced the view that science is just one of many equally valid perspectives on the world, he has often written of social influences on scientific views. Scientific orthodoxy *does* respond to objective evidence about the world, but often slowly, imperfectly and in ways constrained by the prevailing ideology of the times. In short, Dawkins, but not Gould, thinks of science as a unique standard-bearer of enlightenment and rationality. Like Gould, Dawkins accepts the distinction between facts and values: science cannot tell us what to want. But it can tell us what we are.

Part II
Dawkins' World

2   Genes and Gene Lineages

The Selfish Gene begins with a creation myth. Dawkins asks us to imagine a primitive, pre-biotic world – a world in which physical and biochemical processes make available a soup of chemical and physical resources. In this soup, nothing lives, nothing dies and nothing evolves. But then, Something Happens. A *replicator*, by chance, comes into existence. A replicator is a molecule (or any other structure) that in the right environment acts as a template for its own copying. Active replicators have characteristics which influence the rate at which they are copied. Fit replicators get themselves copied often, though their ability to influence their fate will always depend on their environment, too. A replicator that is highly copyworthy in one environment might, for example, be too unstable and hence have very poor prospects in a hotter chemical soup, or one composed of different compounds.

The formation of the first active replicator is a worldshaking event. It is truly something new under the sun, for it introduces natural selection and hence evolution into the world. No copying process is perfect. Hence at some stage, after some number of copyings, the copies of the prime replicator will begin to vary from one another. A population of variants comes into existence. Within the population of variant replicators, some will have better prospects than others. Some will have a higher propensity to be copied. Others will have a lower propensity; they are less stable, or require a less common ingredient in the soup. That creates the conditions for natural selection. For resources are not infinite: the replication of one lineage will have consequences for other lineages. And thus evolution driven by selection begins:

*Competition + variation + replication =
natural selection + evolution.*

The replicators that descend from the original are weeded by natural selection: the variants with features that promote replication will become common; the variants with features that make replication less likely will become rare or extinct.

It would be hard to exaggerate the differences between a world in the first stage of evolution, and our world. Today's genes are made from DNA: specifically, they are sequences of the four bases *adenine*, *guanine*, *cytosine* and *thymine* (usually abbreviated to A, G, C and T) attached to a sugar and phosphate spine. The best-known function of contemporary genes is to code for protein. Indeed, when biologists speak of genes (for instance, when talking of the number of genes carried by particular organisms) they usually have in mind base sequences that specify particular proteins. This specification is implemented by an almost universal code. The base sequence is read in groups of three, each of which (aside from a stop signal) specifies one of twenty amino acids. Hence long base sequences specify amino acid sequences, and such sequences are the 'primary structure' of proteins. The process by which genes produce proteins is indirect, requiring two RNA intermediaries – known as messenger and transfer RNA – and it depends on complex cellular machinery. Moreover many genes are silent: they do not code for protein, and until quite recently they were thought to be inert passengers in organism genomes. It now appears, though, that such genes sometimes play a role in regulating the expression of protein coding genes, helping turn them on and off.

The upshot is that genes, and the gene-to-protein system, are themselves complex products of evolution. The first replicators were certainly not DNA sequences. They may have been RNA sequences (in which *uracil* replaces thymine). RNA does not have a single stable geometric configuration in the way DNA does: DNA's famous double helix formation. This makes RNA's copying more error-prone. But the fact that RNA molecules have many possible physical configurations also makes it possible for RNA to help chemically organise its own replication. In the jargon of chemistry,

RNA but not DNA is a potential catalyst of its own copying. (A catalyst is a molecule that promotes a chemical reaction without being altered by that reaction.) So an 'RNA world' is one possible scenario of evolution's dawning, but even this is very controversial.

So the first replicators were not gene-like DNA sequences. They were not organised into organism-building alliances. Moreover, this was a world of the 'Naked Replicator'. In our world, the genes are replicated, and the organism interacts with the environment both to protect the genes and to secure the resources for their copying. Hence biologists distinguish the *genotype* of an organism (the complement of genes it carries) and the *phenotype* (its developed form, physiology and behaviour). But in this first phase of evolution, the same entity acts both to secure resources and as a template for its own copying. There is replication and interaction with the environment, but there is no specialisation of roles. In a world in which there are no organisms to 'clothe' the replicators – no vehicles – selection will not build better-adapted vehicles. Rather, it will select for molecular properties of the replicators themselves. In Dawkins' memorable phrase, we shall see selection for 'fidelity, fecundity and longevity'.

There is some debate about the aptness of Dawkins' creation myth. Some think life had its origins in a proto-cellular structure which had no chemicals specialised for the role of replication. But if the first

DNA, a double spiral of sugar-phosphate backbones.

Joined across the axis by complementary base-pairs (A to T and C to G)

Genetic information is encoded in the bases' four-letter language.

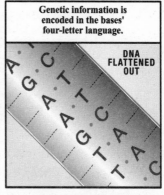

DNA FLATTENED OUT

The code on one strand is transcribed by an enzyme causing a complementary molecule to be made.

MESSENGER RNA (mRNA)

mRNA travels to a ribosome - a double ball of proteins and RNA - to translate its code and to make a protein.

Each triplet of bases on mRNA calls up a complementary triplet on the head of transfer RNA (tRNA).

tRNA

AMINO ACID

Each tRNA carries one of 20 amino acids to the ribosome which detach from tRNA and link together.

← PROTEIN

The growing chain of amino acids twists and folds into a protein.

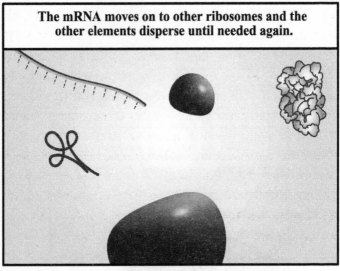

The mRNA moves on to other ribosomes and the other elements disperse until needed again.

Figure 1: Genes in a DNA sequence code for specific proteins. (Source: Steve Jones and Borin Van Loon, *Introducing Genetics*, Cambridge: Icon Books, 2005.)

evolutionary regime was as Dawkins depicts it, there is no doubt that in *that* regime, replicators are the units of selection. For nothing else is undergoing evolutionary change. Perhaps these first replicators are not really living things, but certainly nothing else in this world is at all lifelike. Geologically speaking, this first evolutionary regime cannot have lasted for very long. For shortly after the earth became inhabitable, fossils of bacteria-like organisms appear in the record; these are found in 3.5-billion-year-old rocks from the Pilbara region of Western Australia. So in the space of a few hundred million years, at most, primitive replicators must have combined into physically and functionally united alliances which formed the first cell-like structures. By that stage, as well, genes built from DNA sequences had probably supplanted the ancient replicators whose origin initiated the whole evolutionary process. Life – for bacteria are uncontroversially alive – had by then passed the organism threshold.

The 'invention' of the organism, perhaps even of the simplest cells, changes the character of both evolution and selection. The division of labour has entered the world. For the invention of the organism is the invention of a specialised vehicle for protecting replicators, and for harvesting the resources they need to make new copies of themselves. And some vehicles will be better adapted to their circumstances than others. They will succeed *differentially*. This differential ecological success of vehicles causes the differential replication of the replicators that built those vehicles. If very venomous snakes are more ecologically successful than less toxic variants of the same species, the replicators associated with the venomous variants will be replicated more frequently. In the gene pool of that species, gene lineages associated with the venomous variants will replace lineages associated with less toxic snakes.

So after the organism threshold is crossed, natural selection typically acts directly on organisms and indirectly on replicators. Moreover, it selects replicator *teams*, the whole vehicle-building genome, rather than individual replicators. Life has become essentially cooperative. For if an organism dies, all the replicators in it are destroyed. If it succeeds in reproducing, each replicator in it shares in that success; or, at least, has an equal chance of sharing that success. The genes in an organism typically depend on one another; they share a common fate. In summary, then, passing the organism threshold brings cooperation centrally into the world, and with it a division of labour between work and sex (more strictly, between work and reproduction). Moreover, the effect of a particular replicator on the vehicle carrying it is sensitive to its internal and external environment. Penis-making genes, for example, do not make penises in female vehicles. And they certainly exist in those vehicles. The Y chromosome that male mammals have, and female mammals lack, carries few functional genes. All male mammals inherit from their mother many genes relevant to their male characteristics. Thus a gene's context can make a vast difference to what it does. A toxicity gene in one context may make no difference at all in another. So everyone can agree that in the world of the Naked Replicator, if there ever were such a world, replicators and replicator lineages are the unit of selection. But perhaps the invention of the organism changes the unit of selection.

3 Gene Selection in a World of Organisms

Even after evolution crosses the organism threshold, Dawkins, and before him G.C. Williams, argues that the gene remains the unit of selection. For, as we saw in Chapter 1, selection is cumulative. In *The Blind Watchmaker* and *Climbing Mount Improbable*, Dawkins emphasises the difference between single-step and cumulative selection. In *The Blind Watchmaker*, he points out that if you tried to generate

methinks it is like a weasel

by randomly picking character strings of the right length by repeated trials, you would still be trying at the end of time. The monkey never gets to type even a single sentence of Shakespeare. Cumulative selection transforms the problem. Suppose you make, say, ten random trials and keep the closest, even if it has only a couple of characters in the right place:

qwtxzuifsautysszyaffqyfnmhtb

You then breed from this closest miss, with some error in the copying process. Suppose each daughter string varies from the parent by one character. If so, one of those daughters is likely to have three characters right. Perhaps:

qwtxzuifsaut sszyaefqyfnmhtb

Breed from that one, and so on. Within a manageable number of generations you will reach 'methinks it is like a weasel', though the exact number will depend on luck and copying accuracy. This example is, of course, not a model of *natural* selection; it is more like *artificial* selection. But it does show the great difference in power between single-step selection and cumulative selection. In *Climbing Mount Improbable*, Dawkins reports on models of the evolution of the vertebrate eye from a patch of photosensitive skin, models that suggest this transformation might take only a few million years.

Adaptive evolution requires cumulative selection. In turn, cumulative selection requires persistence. Proto-eyes must be exposed *repeatedly* to selection's scrutiny, if efficient eyes are to evolve. No such eyes will evolve if the proto-eyes of generation 1,000 are significantly different from those of 999 or 1,001. Persistence requires copying; neither individual genes nor individual organisms exist for long. Genes are copied into succeeding generations but organisms are not. Hence the unit of selection must be a lineage of gene copies. QED.

Well, not quite. In *The Nature of Selection*, Elliott Sober has pointed out that persistence need not involve copying. For efficient eyes to evolve, the eyes of generation N+1 must indeed be similar to those of generation N. So the design of eyes must persist across the generations, to serve as a basis for further improvement. But no eye of generation N+1 is a copy of any eye of generation N. Every eye is built from scratch through a complex developmental process. To be sure, genes play an absolutely crucial role in embryological development, but no eye-copying is going on. DNA acts as a template for its own copying. So, perhaps, do some cellular structures as one cell splits into two. But eyes certainly do *not* act as templates for their own replication. Heritability – the similarity of organisms across generations – is essential for cumulative selection. But the heritability of vision does not involve copying eyes themselves. Perhaps, though, it involves copying the eye-making program.

So maybe something has to be copied, for eye design to persist across the generations. One way of putting this idea is that though eyes themselves are not copied, the information needed to make eyes and the rest of the creature is coded in the genes. That information is copied and used. The idea that the genes make up a program or information store is widely endorsed. But it turns out to be surprisingly hard to show that there is a way in which genes and only genes can carry information about a developing organism. Many resources are needed to build an organism. Why think that only some of those – the genes – tell a story about what the organism will be like?

Dawkins takes this challenge very seriously. He accepts, of course, that many resources go into building an organism. An egg is not just a packet of genes; it is a complexly structured biochemical system in its own right. And the egg's environment matters too: for example, in many reptile species the sex of the hatchling depends on its incubation temperature. But (Dawkins argues) though genes may play no unique role in the construction of an organism, they do play a unique role in ensuring that the next generation resembles the current generation. They combine two crucial properties that make adaptive evolution possible. First, they are *difference makers*. Genes never make traits. But often genes make a difference to the traits of their host. DNA sequences at a particular stretch of an organism's chromosome often exist in somewhat different versions; these are known as different alleles of the same gene. Substitute one allele for another on a chromosome and you can often make a predictable difference to a trait. You will see a change in visual acuity, eye colour, body mass or level of intersexual aggression. These differences make the alternative alleles visible to selection: the allele for high acuity can prosper at the expense of the one for lower acuity.

Genes are difference makers, but they are not the only difference makers. Many resources are transmitted between the generations: from ingredients in the cytoplasm of the egg through symbiotic microorganisms (many animals depend on these to digest their food, and their supply is not left to chance) and the social and ecological situation of the parents. Variation in these resources will result in variation in the offspring's phenotype. Differences in parental care, quite obviously, make a difference to offspring condition. However, genes are not just difference makers: they are difference makers in which new variation is preserved; they satisfy the *replicator principle*. Satisfying this condition is essential to cumulative selection. Contrast a new and advantageous genetic variation with, say, a learned improvement in brood care, as a female becomes more vigilant in ensuring that her young are well fed. Both have favourable consequences on the phenotype of the next generation, but the new mothering skill is apt to disappear when the mother dies. When they become mothers, her young (unless there is some special story to be told) will revert to the typical phenotype of the species. Only genetic influence across the generations, because it satisfies the replicator condition, preserves small changes. The mutant mothering gene will make it to the mutants' grandchildren to make them mutants too. These are then available for further improvement, thus making cumulative selection possible.

(Replicator condition)

$$G \Rightarrow G \Rightarrow G \Rightarrow G^* \Rightarrow G^* \Rightarrow G^* \Rightarrow G^* \Rightarrow G^{**} \Rightarrow G^{**} \Rightarrow G^{**}$$

(Reverting to type)

Care $\Rightarrow$ Care $\Rightarrow$ Care $\Rightarrow$ Care * $\Rightarrow$ Care $\Rightarrow$ Care $\Rightarrow$ Care ** $\Rightarrow$ Care

So though parents influence their offspring in many ways, their genetic contributions to their offspring's characteristics play a very special role in evolution: a role which makes the preservation and accumulation of small improvements possible. According to this line of thought, though eyes are not copied, eye-making genes are. Here, battle is again joined. For Gould and his allies deny the existence of eye-making genes. Everyone agrees that evolutionary change is associated with genetic change. After World War II, the myxomatosis virus was introduced to Australia to control the rabbit plague that ravaged the country, and while it initially wiped out vast numbers of rabbits, disease-resistant rabbits soon evolved. When Australian rabbits evolved their resistance to myxomatosis, the gene pool of the Australian rabbit population changed. Evolutionary changes in a population of organisms are tracked by changes in the gene pool of the evolving population. So it is common ground that evolutionary change in a population is correlated with a change in that population's gene pool. But correlation is not the same as causation. The flight of migrating waders to Siberian breeding grounds *correlates* with the start of the Australasian rugby season. But that flight does not *cause* rugby to begin. Similarly, Gould and others reject the idea that the characteristics of genes causally explain evolutionary changes in a population.

At issue is the relationship between genes and the characteristics of organisms they inhabit. Gould would accept the existence of eye-making genes, and selection for such genes, if the eye-maker invariably caused the organism that carried it to develop an eye of a specific kind. In other words, in Gould's mind, gene selection presupposes something like *genetic determinism*. We have to be a little careful in talking about genetic determinism, for no one has ever imagined a gene could make an eye all by itself. Rather, Gould and his allies think that Dawkins is committed to the idea that there is a stable and simple relation between a particular gene and the characteristics of the organism it is in. So, to revert to our rabbits, if there were a specific rabbit gene that always, or almost always, caused a rabbit to be resistant to myxomatosis, then we could say that that gene is a unit of selection, and that it has replicated vigorously in Australia because it makes rabbits myxomatosis resistant. Selection sees through the phenotypic trait of the rabbit – disease resistance – to preserve and copy the gene lineage that is responsible for that trait.

Some genes – those that are *invariant* – do have the same effect on an organism in pretty well every circumstance. They can be quite common in bacteria because the development of the bacterial cell is much simpler than the development of any multi-celled organism. A bacterium acquires an appropriate plasmid – a small parcel of genes – from another bacterium and that acquisition suddenly makes the bacterium and all of its descendants resistant to an antibiotic. For bacteria, the whole problem of differentiation – how different cells move to their adult position and specialise – does not arise. But in multi-celled organisms, an invariant relationship between a gene and the organism it travels in is the exception. When such a relationship is found, it is usually bad news. Most genes with invariant effects cause genetic diseases; they are invariant because they cause something to go badly wrong. Even among genes that cause genetic diseases, a simple relation between a gene and its effect on the organism is the exception rather than the rule. The more usual situation is that most characteristics of organisms are influenced by more than one gene. Malaria resistance in humans is a case in point: it is a consequence of having both the standard haemoglobin-making gene and a variant form of that same gene; the so-called sickle-cell gene. The effect of any particular gene is typically variable and context dependent. In fact, a person with two copies of the sickle-cell gene is in trouble and will probably die of anaemia. If they had instead just one copy of that gene, and one copy of the normal form, they would be fine.

The upshot is that the relationship between genes

and organisms is typically complex and indirect. There is no simple link between genes and traits. No gene makes a trait; few genes are invariably connected with a specific trait. This is most vividly illustrated by the differences between the cells of our bodies. Your liver cells are very different from your neurons in both their structure and their function. Yet they carry exactly the same genes. These basic facts are uncontroversial. However, gene selectionists think that Gould and his allies overstate their importance. Dawkins' views demand that genes have *phenotypic power*. They influence their environment in ways connected to their propensity to be replicated. But that influence can and does depend on their genetic, cellular and ecological environment. The difference-making power of genes can depend on context, so long as these contexts reappear alongside the genes themselves, generation by generation. So, in *The Extended Phenotype*, Dawkins points out that there are 'genes for reading'. Of course, no gene *causes* its bearer to read, come what may. But a gene has phenotypic power over reading if, when substituted for its rivals for the same slot on human chromosomes, the resulting individual is *more likely* to be able to read. Gene selection requires that degree of consistency in a gene's phenotypic effect. But it requires no more: Dawkins, Williams and other gene selectionists are not committed to genetic determinism or anything like it. So far then, a stand-off. Cumulative selection is fundamental to evolution. But that alone

35

does not establish that the fundamental agents in evolution are gene lineages, and the sceptics are right to say so. But equally, Dawkins is right to resist the attempt to hang the albatross of genetic determinism around his neck. In the next chapter, we consider Dawkins' attempt to break this deadlock.

4   Extended Phenotypes and Outlaws

Crossing the organism threshold changes the way selection acts on genes. Before that threshold is crossed, and before genes were assembled into cooperating teams, evolution was a war of all against all. By 3.5 billion years ago, that must already have changed. The genes that built 3.5-billion-year-old cynobacteria were not lone wolves. The relations between gene lineages had already become a balance between competition and cooperation. For no one gene can build a vehicle, not even one as relatively simple as a bacterium. So if the advantages of protection, chemical manufacture and resource harvesting offered by a cell were to be seized, genes had to form alliances of gene complexes. Particular genes in those assemblies earn their place through their ability to influence specific characteristics of their vehicles.

The success of one gene lineage has implications for the success or failure of others. Rabbit gene lineages in Australia are in competition with sheep, kangaroo and wombat gene lineages. There is limited ecological space for grazing animals, and so limited ecological space for their genes. The success of a rabbit-building gene lineage may blight the prospects of gene lineages producing wombat fleas. In a crowded and inter-connected world, the triumphs of one lineage will send causal ripples into many parts of the pool. Flea genes and wombat genes, though, are not irrevocably fated to be competitors. Quite often their fates will be independent of one another. Sometimes, over evolutionary time, different gene lineages in different organisms can become allies. Many fungus genes are allies of tree genes, because there are many associations of mutual benefit between fungi and trees. As Bert Hölldobler and Edward O. Wilson show in their *Ants*, such associations are also very common between ants and trees – many ant genes are allies of tree genes. This dependency can be complete. Bullhorn acacias in Mexico cannot live without their ant allies, *Pseudo-myrmex ferruginea*. They provide them with food (of two kinds) and nest sites. In return, the ants repel plant-eating animals (especially insects) and remove vines and other plants that compete with the acacia for sunshine and nutrition. Equally, genes that travel together in the same organisms are natural allies. For they typically succeed or fail together.

There is a case where competition is inescapable. As we saw in the last chapter, the different *alleles* of a gene are the different DNA sequences within a species that can be found at the same location on a chromo-some. These different versions of the same gene are fated to compete with one another, for alternative alleles in a breeding population are rivals for particular slots on the chromosomes of that population. Triumph for one allele means extinction for the others. Australian magpies are cooperative breeding birds that live in extended families. They are famous for defending their nests with vigour, even against people. The magpie breeding season is a time of terror for cyclists and small children. If a particularly aggressive type of magpie founds new families at a greater rate than others, that differential reproduction will cause a differential replication of the gene or genes responsible for that increased aggression. Copies of the aggression gene will rise relative to alternative alleles that compete for the same slots on the magpie chromosomes. The alternative alleles will drop as a fraction of the population, perhaps to zero. Their lineages will become thinner. So the struggle between rival genes is carried out by collectively constructed organisms that mediate both their interaction with the environment and their further replication.

This is the normal pattern of gene action. Both Dawkins and his opponents can tell a reasonable story about this case. Dawkins' story will be about genes and vehicles. Gould, Sober and others will describe the evolution of magpie aggression in terms of the fitness of individual magpies. But this is not the only way genes lever their way into the next generation. Some genes are loners. 'Outlaw genes' promote their own replication at the expense of the other genes in their organism's genome. Outlaws are relatively uncommon but not unknown (indeed, it seems to be turning out that they are much more common than once supposed). One set of examples are 'sex ratio' distorters. In most circumstances, selection on individual organisms favours a sex ratio of 50/50. But not all genes are equally likely to end up in each sex. Most of the genetic material of complex animals like humans is organised into chromosomes. In our normal cells, we have 46 chromosomes, organised into 23 pairs. These cells are *diploid*. Each gene then exists in two versions, one on each chromosome. Each of these is an allele. These versions can be identical, in which case biologists describe the organism as being *homozygous* at that locus. Or they can be different, in which case the organism is *heterozygous*. When the sex cells (the *gametes*) are formed, this number is halved. Each chromosome pair gives rise to just a single chromosome in the sperm or egg, with its genetic material drawn from the two paired parental chromosomes. So our sex cells have 23 chromosomes. They are *haploid* cells, in contrast to normal cells with 23 chromosome pairs. These haploid cells are formed by a special kind of cell division called *meiosis*.

In most cases, when a 23 chromosome cell is made from a 46 chromosome cell, any particular gene in the parental cell has a 50/50 chance of making it to a sperm or an egg. But not in every case. Some genes are only ever passed on to male offspring; others only ever make it to daughters. Mammals like us have a sex-determining process that depends on the nature of one of the chromosome pairs. A fertilised egg results when sperm and ovum fuse. In our case, each contributes 23 chromosomes and thus builds a new diploid cell with 23 chromosome pairs. A fertilised egg with two X-chromosomes is female; a fertilised egg with an XY pair is male. So sex determination in mammals depends on the male: all female gametes have the X chromosome. No gene on the Y chromosome ends up in a female. So there is selection at the gene-level for any mutation on the Y chromosome that biases the sex ratio towards males, even if that mutation reduces the fitness of the organism with it. Males have an XY sex-specifying chromosome pair, so when they produce sperm, they make some X-carrying (daughter-making) sperm and some Y-carrying (son-making) sperm. So imagine a mutant gene on the Y chromosome that produced fast sperm, sperm more likely to reach the unfertilised egg first. There would be selection at the genetic level in favour of speedy-Y sperm, even if males were in general less fit, being too numerous to have a good chance of finding a mate. Genes often have more than one effect on their carrier. So the speedy-Y gene might have extra, and unfortunate, effects on the male that carries it. But there can be selection for the speedy-Y gene even if speedy-Y males were less fit than other males.

There seems to be at least one case in nature that resembles my imaginary speedy-Y example (in a mosquito), but sex-ratio outlaws that produce female-biased sex ratios are much more common. The package the female contributes is larger and more complex: her egg contains not just a female gamete, but cytoplasm. That complex mix includes extra genes and a whole lot more. So while the evolutionary logic is the same, there are more ways the normal sex ratio can be subverted. As noted, the basic reasoning parallels the male case: a gene that is copied into all an organism's female offspring and only them, and which makes that organism more likely to have female offspring, will have a higher fitness than the other genes. We have genes which are only inherited through the female line. For although most of our genetic material is organised into these 46 chromosomes, not all of it is. *Mitochondria* are energy-generating structures that exist outside the cell nucleus and possess their own genetic material. They are typically inherited *maternally*: your mitochondria are almost always inherited from your mother. The sperm consists of nothing much more than its genetic warhead and a tail (powered by only a few mitochondria that are later discarded), whereas the ovum is a whole cell, stocked with cytoplasm, nucleus and many mitochondria. I am a male, and consequently my mitochondrial genes have no chance of making it to my children. Conversely, my partner's mitochondrial genes are certain to be in any of her children. We have a daughter, but if our daughter had been male, the mitochondrial genes which my partner passed on would be in an evolutionary dead end. So any mutation in those genes that made her more likely to be a daughter-maker would be favoured. There would be selection in favour of any mitochondrial mutation biasing the sex ratio towards females, even if there were deleterious consequences for individual fitness. Such genes are known in plants. They cause plants that are normally capable of producing both pollen and seed to produce only (mitochondria-carrying) seed.

So sex ratio distorters are an example of genes that do not have the same 50/50 chance of replicating as the other genes of an organism. A second class of outlaw genes are 'segregation distortion' genes. When the gametes are formed in a sexually reproducing organism, the normal complement of chromosomes is halved. Normally each allele on each chromosome has a 50/50 chance of being copied into the gamete. Segregation distorter genes bias this lottery in their favour by chemically sabotaging the allele with which they are paired and thus improving their own chances of making it to the gametes. Hence the segregation distorter gene on one chromosome is fitter than its matched gene, its allele, on its paired chromosome. But in increasing their own fitness, segregation distorters often decrease the fitness of the organism carrying them. For organisms with segregation distorting genes facing each other on matched chromosomes are often sterile.

A third form of conflict within the genome is revealed by a phenomenon known as genetic imprinting. Normally both alleles at an active locus contribute to the offspring phenotype (there are exceptions: recessive genes are often genes which have been disabled by mutation, and the distinctive recessive phenotype is a result of gene inactivity). But in some cases, one allele is turned off. While both parents contribute to the offspring genome, the contribution of one is inert. David Haig suggests that this is a result of parents having diverging evolutionary interests. He explores this possibility by retelling a famous biological anecdote. J.B.S. Haldane, a famous controversialist in biology, and one of the triumvirate who created modern evolutionary biology by fusing Darwin's views with modern genetics, is said to have told his students at the pub that he would sacrifice his own life for three brothers. If you think about the genetic basis of such altruism, this makes sense. Imagine that Haldane had carried a gene for such discriminating altruism, perhaps from a new mutation in his mother. Each of his brothers has a 0.5 chance of carrying that same gene, and so his three brothers, on average, would carry 1.5 copies of this gene. So if Haldane's only choices are to live himself, and let all three brothers die, or die, and save those three, that gene will maximise its representation in the next generation, if it can induce Haldane to choose the sacrificial path. The other genes in the alliance that built Haldane are also well served by the same decision, because they too will have copies in those siblings that they project into the next generation. Thirty years later, William Hamilton rediscovered this insight, systematised it, and built it into a theory of cooperative behaviour between relatives.

Haig reasks Haldane's dilemma, but asks whether Haldane should sacrifice himself to save three half-brothers: semi-siblings with whom he shares a mother but no father. Here the genes in Haldane will not agree. Those alleles that came from Haldane's mother will be served by sacrifice. On average, they will do better by Haldane's sacrifice than by his survival. But the paternally derived genes will not want Haldane to sacrifice himself. From their perspective, these half-siblings of Haldane are no relatives at all. A gene in Haldane can promote its own replication directly, by maximising Haldane's chances of having children. Direct replication generates no conflict of interest. The genes in Haldane share a common fate. But they can promote their replication indirectly: by aiding relatives which carry copies of themselves. As Haig points out, indirect replication interests are not the same for every gene in a genome. Haig's maternally derived genes do not have the same relatives as his paternally derived genes. Hence their indirect interests are not served by a common phenotype. Haig argues that genetic imprinting derives from this difference in evolutionary interest. Paternally and maternally derived genes will agree on many aspects of an organism's phenotype. But not all. And disagreement will sometimes take the form of one allele switching the other off. Gene silencing is the result of a struggle for control of offspring phenotypes.

Sometimes, then, we need to think of the genome of an organism not as a harmonious whole but as a gene population. The genes in the population will have overlapping interests. But there is conflict as well, and so there is also differential success; winners and losers. To the extent that we must think of the genes in an organism as a population only partially in alliance with one another, Dawkins' vision is very compelling. To some extent, Dawkins' critics concede this. Outlaws are an uncontroversial case of gene selection. Gould, Sober, Lewontin and many others are sceptical of Dawkins' overall views, but they concede this case to him. But outlaws are not the only case that fits gene selection. In *The Extended Phenotype*, Dawkins argues that there are many examples in which genes reach out into the world to promote their own replication. They have effects, and are visible to selection through these effects. But these are not effects on the organism the gene inhabits. Genes have 'extended phenotypes'. The most vivid examples of extended phenotypic effects involve the action of parasite genes on host bodies. There are many surreal examples of such gene actions. A parasitic barnacle of the *Rhizocephala* group takes over the behaviour of its crab host. After attaching itself to its host, it transforms into a single-celled stage that burrows into the crab, grows and then biochemically castrates and feminises the crab (if it's a male) and subverts the host's brood care behaviours so that these are now supportive of the parasite's own eggs. Elliott Sober and David Wilson in *Unto Others* describe a brainworm that burrows into the brain of an ant, changing its behaviour so that it rests on grass leaves waiting to be eaten by a cow. This is not good news for the ant, but it is for the parasite, for the cow is the brainworm's ultimate host.

Bacteria that are transmitted from parent to offspring are typically beneficial to their hosts. Their fitness depends on their host's success, and many animals (for example, all termites) depend absolutely on their bacterial allies. Termites cannot digest their food without their passengers. But while *Wolbachia* flows from host to offspring, this bacterial lineage travels in the egg but not the sperm. It is passed from an infected host to its *female* offspring. It illustrates the same logic of sex ratio distortion as the mitochondrial variants in plants. Mitochondria are found in specialised structures in cell cytoplasm, and that is where *Wolbachia* live too, and that is why infected males do not pass on their infection through their sperm. It infects various species of invertebrate, and has evolved many dark and unsettling adaptations for turning potential sons into actual daughters. In some species of crustaceans, infected males turn into egg-laying females. In *Hymenoptera* – the lineages of bees and wasps – males are genetically unusual. Unfertilised female eggs develop as males, and as a consequence, males have only a single set of genes. They are diploid. Infected females double the chromosomes in these unfertilised eggs, turning what would have been males into females. In all of these cases, alterations in the host phenotype are due to the adaptive effects of parasite genes. *Wolbachia* and these other examples of extended phenotype effects are carrying difference-making genes. But the differences these genes make are differences to host phenotypes, not parasite phenotypes.

A less brutal example of an extended phenotype is house construction by caddis fly larvae. These larvae typically live on the bottom of streams, and glue together an assortment of debris to form a house in which to live. These houses protect the caddis fly larvae in the same way that a clam shell protects its occupant. But the caddis house is not part of a caddis fly body. It is not part of the organism itself. Structures of this kind can be wonderfully ingenious: Scott Turner's *The Extended Organism* walks us through some impressive and surprising examples. Perhaps the most surprising is the diving bell woven by spiders of the genus *Argyroneta*. These spiders weave an air sack, position it underwater, and use it as an underwater hunting station. But termite mounds are Turner's star example: he shows how these are engineered to exploit the power of wind to drive oxygen into the nest, and suck carbon dioxide out of it. Some termite species live in colonies that can house 1–2 million termites, together with their fungus gardens. So they use a good deal of oxygen, and they produce a good deal of carbon dioxide. They need to exchange gases with their environment efficiently. They do: as well as protecting them, their mound acts as a lung.

Parasite manipulation genes, caddis house-making genes and termite mound-building genes do have effects on the bodies in which they travel. There are many links in the causal chain from one replication of a gene to the next, and the chain begins in the parasite's body. The barnacle produces chemical signals that subvert the host's normal behaviours. The subversion gene directs the production of these chemicals. But the *adaptive effect* of the parasite's gene is its effect on the host's behaviour. Suppose we were to ask: why do these genes exist in the genome of every barnacle in the species? We would answer by describing the feminisation of the crab. For most genes, the path to the future is through their effects on the organism they help build. If a gene contributes to making that organism particularly well adapted, and if it does so in most contexts that the gene finds itself in, it will be replicated frequently. If it typically depresses the fitness of its bearer, it will decline in frequency. In this core case, Dawkins' conception of evolution as a struggle between gene lineages, and Gould's view that selection acts on individual organisms, are roughly equivalent. However, genes have two other replication strategies. Some genes are outlaws, replicating themselves at the expense of others in the same genome. Their replication-enhancing effects are not their effects on the organism in which they ride. Speedy-Y genes do have effects on speedy-Y males: they make it harder for such males to find mates. But that is not the effect which explains why speedy-Y genes would spread in the population. A speedy-Y gene's adaptiveness lies in its local effect on the gamete that carries it.

Extended phenotype genes are not outlaws. The barnacle gene that feminises the crab enhances the prospects of *every* gene in the barnacle genome. But it adapts the environment of the barnacle, rather than the barnacle itself. Its adaptive effect is outside the body it inhabits. Outlaw genes and extended phenotype genes cannot readily be incorporated within a view of evolution that sees selection as acting on individual organisms. Here Dawkins' view of evolution seems to be doing better than that of Gould. A special case of extended phenotype effects concerns social behaviour; the behaviour of groups of animals. To this, we now turn.

5   Selfishness and Selection

Dawkins' most famous book, *The Selfish Gene*, is a response to a pressing evolutionary problem. How could cooperation have evolved? Cooperation is common in the animal kingdom. Many animals co-operate in defending themselves against predators. Musk oxen physically defend themselves as a collective. Many species of jay and crow collectively 'mob' hawks, owls and other dangerous birds. Less spectacularly, but more commonly, many animals warn one another of danger with distinctive calls. A number of carnivores – wolves, African wild dogs, chimps, lions and at least one species of hawk – hunt cooperatively and share their kills. Lionesses tolerate suckling by the cubs of their pride members. Vampire bats that have failed to find blood successfully beg from other bats in their roost. Many species of birds breed cooperatively – parents have 'helpers at the nest', contributing to both the defence of the nest and to feeding the nestlings. White-winged choughs, for example, can only breed with helpers. A single mated pair has no chance of successfully raising fledglings.

So cooperation is far from rare. Yet it poses a puzzle familiar also from human society. For cooperation seems to be *altruistic*, if not in motivation then in its effects. It is true that *everyone* is better off if *everyone* cooperates. Everyone in the tribe is safer if everyone fights bravely in its defence. But if I can get away with it, I am better off still if I quietly withdraw to safety while *everyone else* fights bravely. This is known as the 'temptation to defect'. It has an evolutionary parallel. Think, for example, of a vervet monkey that has just noticed an eagle. Wouldn't it be best off just quietly hiding? Calling can attract the eagle's unwelcome attention. Over time, we would expect selection to cull such traits as warning others about predators, signalling the presence of food, contributing to collective defence and caring for others' young.

What then could explain altruism? There seem to be three possibilities. Altruism might be inadvertent. Animals are not perfectly adapted to their environment. For example, no recognition system is ever perfect. Some risk of error is inevitable. So perhaps the lioness tolerates another's cub at her nipple rather than risk rejecting her own cub by mistake. Tolerating an occasional freeloader would be much less expensive than rejecting her own young. So if there is a chance of such a mistake, caution is reasonable. Yet that caution makes her vulnerable to freeloaders. This possibility might explain a few examples, but it's hard to see how an animal could engage in collective defence or warn others of predators by mistake. No error hypothesis can explain all instances of cooperation.

A second idea is that cooperation is not altruistic: agents benefit themselves by cooperating with others. The idea here is to explain away the appearance of altruism. This possibility is central to contemporary debates on this issue. Bernd Heinrich's *Ravens in Winter* explores one such case. Heinrich was puzzled by the fact that when an individual raven found a carcass – a rich source of food – it seemed to advertise its find rather than attempt to monopolise it. Why would a raven do such a thing? It turned out that ravens who call at the sight of large carcasses are not acting altruistically at all. Adult ravens hold territories, whereas juveniles do not. The ravens that call when they discover a carcass have no territories of their own. By themselves they will be chased off by the territory owner and end up with nothing or almost nothing. So they call and in doing so they recruit others. The recruits swamp the owner's territory defence. The calling ravens will have to share with those they recruit, but they will still get some of the windfall.

In pursuing this line of thought, evolutionary biologists have been particularly interested in two ideas. One is the idea that cooperation involves trading benefits. If two or more animals can secure some resource by cooperating that neither could secure individually, individual selection could promote joint action. Social predators – such as wolves and African wild dogs – take and then share prey that no individual could kill by itself. It is in each dog's interest to act with the others, so long as the individual's share of the joint carcass is more valuable than any prey it could catch itself. Reciprocal altruism takes this unproblematic form of cooperation as its model and extends it to cases in which the partners do not reap their reward simultaneously. Each is better off trading than not trading, and each animal in the trade is vigilant to ensure it is not cheated. This view of cooperation has been strengthened in the last decade and a half by the work of Robert Axelrod, who has shown that the strategy of 'tit-for-tat' can pay off in many situations. Tit-for-tat is the rule of cooperating on the first interaction with another animal, and thereafter doing what it did last time. So if your partner cheated on the first interaction, you should cheat on the second. If your partner cooperated, you cooperate. The best-known biological example is from the life of vampire bats. They share blood. These bats die unless they feed every couple of days, and hunting failure is quite common. So reciprocation is an essential element in vampire bat life. Successful bats share with those who fail. But bats that give are bats that receive. Tit-for-tat seems to be a good model of cooperation between pairs of agents who interact often. But it may not explain cooperation in larger groups. In the face of a defection, the individuals in a larger group have only two choices. They can defect themselves next time, punishing everyone, not just the cheat. Or they can ignore the initial cheating. Neither response will keep cooperation alive.

The problem of social interaction, especially the problem of cooperation, has driven the development of many of the most important new ideas in evolution. One particularly important idea is evolutionary game theory. When an animal interacts with its physical environment, its fitness does not usually depend on the other members of its population. A gene in a tiger that improves visual acuity or metabolic efficiency will benefit that tiger, whatever other tigers do. Such traits are beneficial independently of their frequency in the population. But the fitness effects of social traits often depend on their frequency. Even if wolves would be better off hunting cooperatively than going it alone, a wolf with the disposition to cooperate will gain no benefit thereby unless others have it too. Other traits have the reverse dynamic. In a population of cooperators, the rare cheat does very well. John Maynard Smith developed a famous model to show that social traits, including cooperative ones, often have equilibrium frequencies at which two opposed traits both persist in the population. Maynard Smith imagined a population that had no real contests over important resources; for example, a nesting hole. If two birds both wanted the same hole, both would bluff for a while, and then one would give up. No bird would actually fight for a hole. These birds follow a 'dove' strategy in their interactions. This population would be vulnerable to an invasion by a bird playing 'hawk', actually attacking to fight for a hole. In a population playing dove, hawks do superbly. They always get their hole, and they never have to pay the cost of actually fighting. But as the frequency of hawks rises in the population, the cost of being a hawk rises too. For now they begin to meet other hawks, not doves, at the holes. They do not always get their hole, and they have to bear the costs of fighting. Maynard Smith showed that unless holes are very valuable, or the risks of fighting are low, there will be an equilibrium frequency of both hawks and doves in the population (or a frequency with which each bird sometimes acts like a hawk, and sometimes a dove). This frequency is evolutionarily stable. Axelrod's work shows that, in important circumstances, tit-for-tat is an evolutionarily stable strategy. A population all following this strategy cannot be invaded by any mutant following an alternative strategy.

A second strategy for explaining away altruism is based on another unproblematic form of cooperation. Many animals aid their own offspring. In doing so, they project their own genes into the future, for their young bear their genes. But that is true not just of an animal's direct offspring. An animal's kin, especially its close kin, carries copies of its distinctive genes. Sometimes an animal can best project its genes into the future by aiding kin. Hence Haldane's promise (noted in the last chapter) to sacrifice himself for three or more brothers. Behaviours that evolve through benefit to kin are known as kin-selected effects. The measure of fitness which includes these indirect effects is known as 'inclusive fitness'. Kin selection probably explains many cases of cooperation within family units. But, clearly, it cannot explain cooperation between agents who are distantly related, or not related at all. And there are many such cases. Vervet monkeys, for example, do not live only with relatives, and yet they warn one another of predators.

A third idea accepts that cooperation is often individualistically altruistic: aiding others does have a price for the individual. But that aid improves the fitness of the collective to which both individuals belong. Those collectives do better than rival collectives of individuals who fail to cooperate. The idea then is to explain cooperation as a result of selection on the population of collectives. In some species of baboon, the adult males defend the troops of which they are a part. On this line of thought, the collective is itself a vehicle of selection; it is a *super-organism*. The cooperative baboon troop is more likely to survive and is more likely to found new troops like itself than one composed of baboons following the maxim of 'every baboon for itself'. According to this suggestion, the population of baboon troops is a salient level of biological organisation. It forms a population of mini-populations (the troops) where these mini-populations compete with differential success.

Gene selection was born when G.C. Williams, in *Adaptation and Natural Selection*, defended the idea that altruism is an illusion against such group selective theories of altruism. Williams argued that selection at the level of groups would in almost every case be undermined by individual selection for defection pushing in the opposite direction. The evolutionary temptation to defect is too great to allow real altruism to evolve. A selfish individual in an altruistic group would always be fitter than his fellows – so the altruism of the group would be eroded from within. Moreover, wolf packs and similar groups last longer than the individuals within them. The group's lifespan is longer than the individual lifespans. So the clock of *individual* selection runs faster and hence more powerfully than that of *group* selection.

In this respect, *The Selfish Gene* follows in Williams' footsteps. For Williams and Dawkins, cooperation is real but altruism is not. So they aimed to explain away the appearance of altruism. They are both very sceptical about group selection; of the idea that groups are super-organisms. Groups are not themselves adapted. Rather, they are shifting ensembles of individual organisms. A fast herd of horses is just a herd of fast horses: the adaptation – speed – is a trait of the individual horses in the herd, not of the herd itself. Yet group selection presupposes that the herd itself is adapted.

There have been very important developments in thinking about cooperation since *Adaptation and Natural Selection* and *The Selfish Gene* were first published. One is the realisation that the evolution of cooperation is a truly fundamental feature of biology; cooperation is not restricted to a few odd animals and possible only under special conditions. For the evolution of the organism itself is an instance of co-operation. John Maynard Smith and Eörs Szathmáry showed this very vividly in their brilliant and pro-vocative *The Major Transitions in Evolution*. For them, the evolution of biological complexity is essen-tially the evolution of increasing cooperation. The evolution of gene teams, joint phenotype building, joint and fair replication, and sex cell formation all posed cooperation and defection problems. In sex cell formation, for example, only half the genes in the ancestor cells, each of which has two copies of each chromosome, make it to the gametes, each of which has only one. Despite the existence of some outlaws, the lottery is and must be typically fair (otherwise sexual reproduction would be evolutionarily unstable), but each gene would improve its fitness if it could bias the lottery in its own favour. This situation is typical. The fitness of every replicator is boosted if all cooperate, say, in building a cell before replicating. But individual replicators had evolutionary temp-tations to defect, to become outlaws. The existence of segregation distorters and other outlaws shows that in the evolution of the organism, this defection problem has not been fully solved. Cancers are cells that have become outlaws. But the fact that organisms have evolved shows cooperation can evolve in a fair range of circumstances.

Moreover, it strongly suggests that the evolution of cooperation sometimes involves selection acting at two levels: on individual agents and on the collectives individual agents form. For consider one of these transitions in which a new form of biological organisation evolves when previously independent players fuse into a single more complex individual. The evolution of gene teams, eukaryotic cells, multi-celled organisms and (arguably) the social insects are all examples of such transitions. At the beginning of these transitions, selection acted on (say) individual cells or individual proto-ants. At the end of it, selection primarily acts on collectives: the ant nest, the multi-celled organism. But if these transitions are gradual (as everyone supposes) through much of the transition, selection must have acted on both ants and nests; cells and cell collectives.

It has also become clear that Dawkins' gene selection is compatible with the idea that cooperation evolves by selection on collectives, not individual cooperators. Dawkins agrees that organisms play a central role in evolution. They are vehicles of selection: their success determines the proliferation of gene lineages. The most prominent current defenders of group selection, David Wilson and Elliott Sober, point out that group selection is a claim about vehicles. The group selection theorist can agree that at its most fundamental level, evolutionary history is the history of the success and failure of competing gene lineages. But they claim that some rival gene lineages compete by coding for the characteristics of groups. Genes can make groups differ in their ecological success. Those differences will matter. If there are baboon genes which induce baboon groups to defend the whole group aggressively against threats from leopards, that is just another example of a gene with an extended phenotype. So, in embracing gene selection, Dawkins need not reject group selection. This much is now uncontroversial.

Finally, new and more plausible models of group selection have been developed. Williams and Dawkins confronted and rejected the idea that wolf packs and baboon groups were super-organisms. A wolf pack, on this picture, is an organised, co-adapted repro-ducing creature (that happens to be built from smaller creatures, just as we are built from cells). Packs do not give birth to wolves (wolves do that), but to new packs. So when we measure the fitness of a pack, we measure the rate at which it founds new packs. Since wolves breed faster than packs, Dawkins, Williams and Maynard-Smith all conclude that the cancer of individual lupine selfishness would eat away pack cooperation faster than pack-foundation by cooperat-ing packs could build pack cooperation. Of course, you might think that cooperative wolves would force those tempted to be selfish to behave: they would punish failures to cooperate, and hence suppress the cancer of cheating. But even for a wolf, biting another wolf is dangerous. Punishment to force good beha-viour is itself an act of cooperation: so the existence of punishment (where it does exist) is just as puzzling as any other form of acting for the good of the group. So scepticism about group selection remained. As the sceptics pointed out, while it was not strictly impossible for cooperation to evolve by selection on wolf packs rather than wolves, cranking through the numbers seemed to show that it would require very special conditions. The arrival of a selfish wolf into a cooperative pack would have to be very rare, or the profit from cheating would have to be surprisingly low.

However, in the last decade or so, David Wilson (lately with Elliot Sober) has developed an alternative version of group selection in which the profit of cooperation shows up in the rate at which cooperative packs churn out new wolves rather than the rate they form new packs. Cooperative packs (Wilson suggests) are more efficient machines for producing wolves than less cooperative packs, and most of the wolves so produced will be cooperators, and hence will be apt to join cooperative packs themselves. On this way of formulating the issues, the clock of group selection runs at the same pace as that of individual selection, and hence there is no reason to assume that selection for individual selfishness (for within mixed packs of cooperators and cheats, cheats still do better) will trump the effect of cooperative packs contributing more to the next wolf generation.

On this issue the views of Dawkins and Gould have converged to some extent. It is true that Dawkins remains sceptical of group selection, even in its modified form, whereas Gould, in *The Structure of Evolutionary Theory*, endorses this model of co-operation and its evolution. But while Gould is very interested in the idea that there is selection on the collectives individual organisms form, and not just on those individuals themselves, the collective he is most interested in is that of the species. He is more interested in species selection than group selection. While showing some caution, Gould is rather persuaded by the idea that species differ both in characteristics that make them vulnerable to extinction, and in charac-teristics that make them evolutionarily fecund. For example, species whose gene pools contain plenty of variability are, other things being equal, more resilient in the face of environmental change than species with relatively little variability. The same is true of species with broad geographic ranges. Broad-ranging species are more buffered against change and hence less likely to go extinct than those with narrow habitat tolerances.

The species selection hypothesis is both extra-ordinarily intriguing and deeply problematic, as we shall see in Part III. But it does not pose a defection problem. Group selection and selection on individual organisms are candidate explanations of traits of the same kind: warning calls, food sharing, joint defence and the like. That is why selective forces can act in the opposite direction. The fact that group selection favours, say, collective defence and that individual selection selects against it opens the door to the problem of defection and hence to the possibility that individual selection will be more powerful than group selection. This problem does not arise for species selection. For the traits relevant to species selection are not traits of individual organisms at all. Consider the candidates. They include such characteristics as geographic range, gene pool heterogeneity and the like. These are properties of populations not individuals. So the main problem Williams and Dawkins have urged against group selection, the defection problem, does not arise for the version of high-level selection Gould has been exploring. Moreover, Dawkins' own views have edged towards that of Gould. In *Climbing Mount Improbable* (in chapter 7) and in *The Ancestor's Tale* (in chapter 39), he discusses the evolution of evolvability itself. Some lineages of animals are more *evolvable* than others – something about the basic organisation of the animal makes it easier to generate evolutionary change. In this connection, Dawkins discusses the evolution of body segmentation. Arthropods are animals with exoskeletons and jointed, segmented bodies. Spiders, crabs and insects are all arthropods. Perhaps it is no accident that the segmented-bodied arthropods are far and away the most diverse lineage of animals. For once segmentation has been invented, natural selection can specialise segments to new roles. Thus their limbs have often been converted into feelers and various other specialised biological machines. Dawkins suspects that the rich radiation of segmented animals is explained by some kind of high-level selection for evolvability, albeit one that is not opposed by individual-level selection. Many important questions about selection remain open. But on these issues, the divisions between Dawkins and Gould are less sharp than they once were.

6   Selection and Adaptation

I have just suggested that the fissure lines between Dawkins and Gould on high-level selection are not as wide as they once were. Despite the heat of some recent rhetoric, the same is true of the role of selection in generating evolutionary change. In 1989, Gould collaborated with Richard Lewontin in a famous critique of evolutionary biology, arguing that, as it was then practised, biology was *adaptationist*. It was not so very obvious what adaptationism was. But two aspects of this sin were clear. Evolutionary biologists were too ready to assume that characteristics of an organism were shaped by natural selection for some function. And they were too easily satisfied that they had discovered that function. Most evolutionary biologists would agree that the 1989 paper had a salutary effect, stimulating the development of new ways to test evolutionary and selective hypotheses. One method has been to turn a hypothesis into a formal mathematical model, one making quantitative and measurable predictions about a population. As we noted in Chapter 4, wasps, bees and other social insects have an unusual genetic system: males develop from unfertilised eggs, and, like our sex cells, they have only one set of chromosomes. Queens and workers are female, and develop from fertilised eggs. It follows that if the queen has mated only once, the sister workers in a nest are more closely related to one another than they are to their mother. As with us, they have one chance in two of carrying any particular one of their mother's genes. But they share, on average, three out of four genes with their sister. For they all get the same set of genes from their father, since he has only one set to give. His sperm are all alike. So the sisters share all their paternal genes, and on average half their maternal genes, making three out of four in total. These facts lead to different expectations about the colony sex ratio if it is queen-controlled than if it is worker-controlled. The queen wants more sons than do the workers. So formal models can then be constructed from and compared with actual data to test for worker-versus-queen control.

A second approach has been to develop explicit comparative methods: methods that compare the species under study with its relatives. The idea is to distinguish traits which are adaptations to current circumstances from traits which are inherited from the species' ancestors, by looking at its relatives. Suppose we wondered why the golden-shouldered parrot lays its eggs in holes in termite mounds. The fact that this parrot lays its eggs in a hole, rather than in a nest that it builds, is probably not an adaptation to its specific circumstances. For all parrots lay eggs in holes. Nesting in a hole is probably a trait that the golden-shouldered parrot inherited from its ancestors. There is a quite serious dispute about how ancient the parrots are. It used to be thought that though the first birds were the dinosaur's contemporaries, the modern bird lineages (including the parrots) evolved after – but not long after – the mass extinction in which the dinosaurs died. That is reckoned to have been about 65 million years ago. There have been recent, revisionary ideas that push the origin of the modern birds back into the time of the dinosaurs. But even if those are wrong, parrots have been laying their eggs in holes for upwards of 50 million years, and so the golden-shouldered parrot's failure to make a nest is no response to its current circumstances. But most parrots lay eggs in natural holes in trees rather than excavating holes in ant nests. Only the golden-shouldered parrot and a few close relatives, all living in rather treeless grasslands, use termite mounds. So this probably is an adaptation to their specific ecological circumstances. Both these techniques are still being developed, but there is no doubt that evolutionary biologists have responded constructively to the Gould–Lewontin challenge.

Even so, whatever adaptationism is, Gould thinks that it is still alive, and his *Structure of Biological Theory* includes a long chapter on its historical development. In his notorious review of Dennett, Gould accuses him of representing an 'ultra-Darwinian' strand of evolutionary thinking; of believing that just about every characteristic of every organism is shaped by natural selection. There are important differences between Dennett and Dawkins on the one hand and Gould on the other on the role of selection in driving evolutionary change. But this is not one of them. Everyone accepts that many characteristics of organisms are not the direct result of selection. Consider, for example, the male king parrot, which is a brilliant red. It is probably red as a direct result of sexual selection: females prefer red males. But nothing like that is true of the colour of the parrot's blood. It is a by-product of selection for blood's real function, that of carrying oxygen to the tissues. Such examples can be endlessly multiplied. Some traits of organisms result from chance fixations of neutral characteristics; indeed, that seems very likely to be true of many of our genetic characteristics. Some characteristics are the result of an inheritance from distant ancestors that is now entrenched in the way an organism develops. As Gould points out in one of his engaging essays, that is very likely why we have five fingers and toes. Some traits are adaptive vestiges. Many cave-dwelling species have non-functional eyes that are vestiges of the functioning eyes of their sighted ancestors. Some characteristics of organisms are by-products of selection for some other characteristic. Human female reproductive anatomy is jerry-built as a consequence of our adaptation for bipedalism.

None of these general truths are at all controversial,

though their application to particular cases may be. Nor is there disagreement between Gould and Dawkins on core cases. For example, Dawkins begins *The Blind Watchmaker* with a discussion of bat echolocation. The fossil record of bats is not particularly rich: there are Eocene fossils, but these are of bats that have already evolved echolocation. Even so, no one doubts that this system evolved in bats as a means for them to locate themselves in space and to locate their prey. The echolocation system is complex and integrated. It powers a very distinctive type of behaviour that is central to the life history of bats, and it supports only that type of behaviour. These facts allow us to both identify echolocation as an adaptation and to identify what echolocation is for.

Everyone agrees that bat echolocation is an adaptation. As Gould says, 'eyes are for seeing and feet are for moving' (*New York Review of Books*, 12 June 1997). In his *Structure of Evolutionary Theory*, he writes 'I do not deny the wonder, or the powerful importance, of organised adaptive complexity. I recognise that we know no mechanism for the origin of such organismal features other than conventional natural selection' (p. 710). But once we move away from these uncontroversial cases, identifying adaptation is difficult and controversial. This fact is of particular interest to Gould because the application of evolutionary theory to human behaviour has often involved a shift from these clear and unmistakable cases to much more doubtful ones. For instance, in

71

their notorious *A Natural History of Rape: Biological Bases of Sexual Coercion*, Randy Thornhill and Craig Palmer argued that sexual jealousy and rape are adaptations. But neither show the adaptive complexity that so uncontroversially marks echolocation as an adaptation. What would show, say, that jealous rage is an adaptation rather than an unfortunate side effect of our emotional repertoire? An optimality analysis would have to estimate the (very real) fitness costs of sexual violence and its supposed benefits. This would be especially challenging, for both costs and benefits must be relativised to different environments. We would then need to derive a prediction about the optimal rate of violence (again, relativised to the differing environments in which humans find themselves), and compare our predicted rate to a measured rate. A close fit would support the conjecture. Nothing like this is to be found in Thornhill and Palmer's work.

Adaptationist claims of this ilk, perhaps rightly, make Gould's blood boil. But here he has no direct argument with Dawkins. Dawkins, unlike Gould, is confident that some human behaviour patterns are adaptations; a confidence he makes clear in an assertive review of Steven Rose, Leon J. Kamin and Richard Lewontin's *Not In Our Genes*. But *Climbing Mount Improbable* and *The Blind Watchmaker* focus on the uncontroversial cases. For Dawkins takes the explanation of adaptive complexity to be the central task of evolutionary biology. He does so precisely because such systems are highly improbable, and hence can be explained only by natural selection. These books focus on clear cases. On these, he and Gould are in agreement.

There are, however, important disagreements. One is over the relative role of selection and variation. Selection acts only on variation generated in a lineage. The developmental biology of a lineage determines the range of variation. That developmental biology is the result of that lineage's evolutionary history. So the variation that is available to selection in a lineage is determined by its history: its history *constrains* its future evolutionary possibilities. Perhaps a chimpanzee with a powerful and prehensile tail would be fitter than chimps as we find them. For it would be well suited both to life in the trees and on the ground, harvesting the best of both worlds. Even so, if no tailed variants were thrown up in ancestral chimp populations, selection cannot make such a chimp. The evolutionary trajectory of a population is hostage both to selection and to the supply of variation. In his *Biased Embryos and Evolution*, Wallace Arthur gives some engaging examples of surprising gaps in nature from his own studies of centipedes. Geophilomorphian centipedes are a family of 1,000 or so species. As with all centipedes, their trunks are divided into segments, on each of which is a pair of legs. In this family, the number of segments varies wildly: from 29 to 191. But only in odd numbers; there are no centipedes from this family that have, say, 66 pairs of legs. Lithobiomorphian centipedes are another family, again of

1,000 or so separate species. They all have the same number of segments: fifteen. Given the variation in segment number in centipedes as a whole, it seems unlikely that there is some selective advantage to having an odd number of leg pairs, or of having exactly fifteen leg pairs. On the face of it, our missing centipedes seem most likely to be the result of a constraint on the supply of variation to centipede evolution. No one doubts this potential role of constraints on variation on the evolutionary trajectory of a lineage. If the developmental biology of Lithobiomorphian centipedes cannot build a 30-segment centipede, no such animal can be selected. But Wallace points out that constraints do not have to be absolute to influence developmental trajectories. Bias is enough. Imagine, for example, a population of ringtail possums on an offshore island, unbothered by predators. Suppose that they average two kilograms each. Unfortunately, cats arrive, and they face new threats. If they were larger, they would do fine: a three-kilo possum is too large for these cats. If they were smaller they would be fine: a one-kilo possum can hide in places where cats cannot go. Which will evolve? The trajectory of the possum population is likely to be determined by biases in the supply of variation: whether the current population is more likely to throw up small possums than large ones.

This issue about the supply of variation has led to a major debate in evolutionary biology about the relative role of the supply of variation and selection in explaining evolutionary change. We can explore this debate through one of Dawkins' own examples in *Climbing Mount Improbable*: the 'Museum of All Possible Shells'. Minor details aside, it turns out that shells vary in just three ways: the rates at which they uncoil in one plane (their 'flare'), the rate at which they rise above that plane (their 'spire'), and the rate at which their tube expands (their 'verm'). This makes it possible to represent the space of all possible shell shapes as a cube, where each dimension of that cube corresponds to one of the three ways shells vary from one another. Thus any point in the cube represents a possible shell: a shell made by expanding in a plane at a given rate; expanding its tube at a specific rate; and rising above the plane at a specific rate. Most of these possible shells do not exist, and to the best of our knowledge never have existed. A large chunk of that cube is unoccupied. What explains the missing shells? Are these missing variants impossible for shell lineages to generate? Have shell-building lineages inherited insufficient variation to build these missing shells? Perhaps, instead, they have been edited out by selection; they are too expensive to build, or too unwieldy, or too fragile. These questions remain unanswered: both about shells in particular and about all the many apparently possible plants and animals that have never existed. Why are there no centaurs? Perhaps they would be too expensive to run, or too subject to back pain. But perhaps six-limbed mammals have simply never been available for selection.

Figure 2: Raup's cube representing the space of all possible shell shapes. Regions of the cube in which real-life shells can be found are shaded. The unshaded regions house theoretically conceivable shells that do not actually exist. (Source: David M. Raup, in Raup and Stanley's *Principles of Paleontology*, London: W.H. Freeman, 1979.)

Both Dawkins and Gould recognise the fundamental importance of these issues. In his *Art of the Developable*, Dawkins notes that the central gap in contemporary evolutionary theory is 'its lack of a complete theory of variation, of the origin of candidates for evolution' (p. 202). Development has been even more fundamental to Gould's work. His *Ontogeny and Phylogeny* is regarded by many as a foundation of modern evolutionary developmental biology, and his interest in the interrelation of development, variation and evolution has never ebbed. However, though both have been interested in development, their emphases have been different. Dawkins has not given the idea of developmental constraints, the idea that development limits evolutionary trajectories, much weight. His guess is that, over the long run, the space of evolutionary possibilities open to a lineage is rich. Hence the history of the lineage is largely determined by selection making some of these possibilities actual. Selection determines, for instance, that actual mussel shells are strong, thick and low. Dawkins' interest has been in the role of developmental revolutions in enhancing the possibilities open to a lineage. In *Climbing Mount Improbable*, he discusses the evolution of segmentation in this context: once segmentation is invented, segments can be duplicated and they can diverge from one another, making the evolution of many specialised capacities possible. He returns to similar themes at the end of *The Ancestor's Tale*: major transitions in evolution are developmental transitions, transitions which make new variants possible, and hence new adaptive complexes possible.

Gould, on the other hand, is inclined to bet that the array of possibilities open to a lineage is tightly restricted, often to minor variants of its current state. Hence the history of a lineage is largely shaped by the events that set the envelope of possibilities; for instance, the events that determined that vertebrates have at most four limbs. One of Gould's fundamental interests has been in conservative patterns in animal evolution. As he sees it, the history of animal life since the base of the Cambrian is a history of morphological stability, a stability probably explained by constraints on the supply of variation to selection. So especially in his earlier work (in particular, in *Wonderful Life*), he tended to think of the supply of variation as a brake on evolutionary change, though in *The Structure of Evolutionary Theory* he is careful to point out that it can enhance possibilities for change as well. So while both Dawkins and Gould recognise the central role of developmental biology in an explanation of evolutionary change, they make different bets as to what that role will be. Gould but not Dawkins thinks that one of those roles is as a brake, damping down the possibilities of change. To this difference is added another. For Dawkins, the central problem of evolutionary biology is the explanation of adaptive complexity. That is not Gould's conception of the field.

He has spent a large fraction of his palaeontological career arguing for the existence of large-scale patterns in the history of life, patterns not explained by natural selection. A further disagreement concerns the existence and importance of these patterns. To this, we now turn.

Part III
The View from Harvard

7   Local Process, Global Change?

Gould sees himself as having two fundamental disagreements over selection with Dawkins and others of
a similar mindset. One is about evolutionary changes
within a species. Microevolution is the branch of
evolutionary biology concerned with evolutionary
changes taking place within a species; changes that
occur at a scale we can observe. Gould thinks that
Dawkins and other gene selectionists misrepresent
the role of genes in microevolution. Gene differences
do not cause evolutionary changes in populations,
they register those changes. It is important to keep
track of genetic changes, because they indicate the
presence of evolutionary changes in population, and
they can be identified and counted. We can and should
use gene changes as symptoms of evolutionary change,
but we should not confuse symptoms with causes.

Moreover, Gould thinks that evolutionary biologists
too often neglect non-selective possibilities when
they formulate and test their hypotheses about
microevolutionary change. In his critique of human

sociobiology, this has probably been his main beef. For example, E.O. Wilson has argued that males and females differ predictably both in their sexual behaviour and in their behaviour towards children. Men are more apt to be promiscuous than women and are also less inclined to pour all their resources into a single monogamous partnership. Gould is sceptical even of these sociological claims about the way we actually behave, and with reason: Sarah Blaffer Hrdy in *Mother Nature* is just one evolutionary biologist to give a much more nuanced account of male and female reproductive roles. But Gould is even more sceptical about the adaptationist explanations offered for these activities. Suppose Wilson is right, and that, as a rule, men and women tend to differ in these ways. These differences may be no adaptation at all. Instead, they might be a vestige of sex differences inherited from our hominid ancestors. One of the most famous of all hominid fossils is 'Lucy'; an extraordinarily complete fossil of a female *Australopithecus afarensis*. This hominid species lived about 3 million years ago and was highly sexually dimorphic; that is, the males were much larger and more robust than the females. Perhaps the differences between men and women are just a reduced residue of this much greater original difference. Gould thinks sociobiology has systematically neglected these non-adaptive possibilities, an argument he presses with great vigour in the second of his *New York Review of Books* contributions.

Despite his heat on these issues, this has been the

lesser of Gould's concerns. His main target is a view he calls 'extrapolationism'. Extrapolationism concerns the relationship between evolutionary processes that take place within a species and those of the large-scale history of life. Most species are fragmented into local populations, living in environments that vary to some extent from one another. In some species, this fragmentation and isolation can be quite extreme. Barn owls live on every continent but Antarctica, and are scattered through environments that differ in climate, vegetation, competition and predation. They are close to one end of a continuum that runs from very widely distributed and unspecialised species like this owl, to species which live only in one tiny corner of the world. But few species consist of a single homogenous population. Sometimes the local populations into which species are divided are fully isolated; they are cut off in 'islands' of suitable habitat. But, for the most part, there is some migration in and out. Even so, the members of these local populations mostly interact with other locals, both in competition and reproduction. So natural selection takes place within these fragments. Since the different populations are different samples of the variability of the whole species, and because the environments vary, different populations of the same species often diverge from one another, though that divergence is often temporary. It breaks down when populations rejoin.

How do events at this scale relate to the large-scale history of life? How do changes in local populations over a few generations relate to the evolution of species and lineages of species documented in the fossil record? Gould argues that mainstream evolutionary biology has accepted an extrapolationist view. Indeed, he thinks this view dates back to Darwin himself. In this view, the evolution of species lineages is nothing more than an aggregation of events at the scale of local populations. Major changes are minor changes added up over many generations. Evolutionary patterns are generated by the processes documented in local populations. One of Darwin's greatest insights was of the transforming power of a long series of events, each of which is individually minor. This was originally a geological insight: great cliffs are both laid down and eroded away a few grains at a time. Darwin extended this vision of the transformational power of small events long repeated from rock and earth to shell and bone. Just as landscapes are transformed by a long series of mundane events, so are lineages. While not disputing the importance and power of this insight, Gould argues that it is not the whole truth. Indeed, it is not much of an exaggeration to say that Gould's professional life has been one long campaign against the idea that this history of life is nothing but the long, long accumulation of local events. To begin with, I will review four highlights:

(i) Gould's first famous contribution to evolutionary thinking came in 1972. With Niles Eldredge, he developed the theory of 'punctuated equilibrium', a view of the typical life history of species. According to

Gould, species do not gradually evolve into new species. *Homo habilis* did not gradually, imperceptibly turn into *Homo erectus*. Rather, new species typically arise by a split in a parental species followed by rapid speciation of one or both of the fragments. The typical life history of a species involves its geologically rapid formation. New species typically appear in the fossil record already fully differentiated from their parent species. Their distinctive characteristics are already present in the earliest fossils, rather than gradually emerging over the species' life history. Once a new species appears, it usually undergoes no further evolutionary change until it goes extinct, or until it splits into daughter species. Gould argues that the punctuated equilibrium pattern challenges extrapolationism. Extrapolationists ought to expect gradual change in a species. Extrapolationism predicts the gradual evolutionary accommodation of a species, as it slowly changes to suit new environments (for that is how we see local populations respond). Moreover, if species life histories do have this pattern of rapid formation followed by stability, we need a new explanation of evolutionary trends. Hominid evolution is a classic example of an evolutionary trend: over hominid history, there has been a marked increase in relative brain size. But if such hominid species as *Homo habilis* or *Homo erectus* showed no significant evolutionary change after they originated, this trend cannot be produced by a slow growth of relative brain size over the lifetime of a species. Trends must be the result,

Gould concludes, of species sorting. Species with relatively larger brains must have been more likely to appear, or to survive.

(ii) In many of his *Natural History* writings, Gould has argued that mass extinctions have had a profound effect on the history of life, an idea to which he returns in the final chapter of *The Structure of Evolutionary Theory*. Gould was an early supporter of the idea that an asteroid impact caused the Cretaceous/Tertiary extinction, the one that saw the end of the pterosaurs, large marine reptiles and the non-avian dinosaurs 65 million years ago. If a large impact caused those extinctions, then they were sudden, even in ecological terms. Moreover, as Gould reads the record, the big rock did not just finish off a doomed lineage. Had it missed the earth, the dinosaurs might still be dominating terrestrial ecosystems, whales would never have had a chance and mammals might still be rat-sized insect-eaters skulking in the dark. Mass extinctions do not strike at random. Some types of species are more vulnerable than others. But the extent to which the organisms that make up a species are adapted is irrelevant to the survival prospects of that species. For adaptation is adaptation to a specific environment. After the asteroid landed next door, the environment to which Mexican dinosaurs were adapted ceased to exist. Mass extinctions are caused by events which disrupt those environments catastrophically. They suddenly change the rules of the game. Since those changes are sudden and severe, selection is powerless to adapt organisms to changed circumstances. Many species of dinosaur were very likely superbly adapted to their habitats, but that is irrelevant if those habitats were destroyed. The properties that are visible to selection and evolution in local populations – the extent to which an organism is suited to life here and now – are irrelevant to the prospects for survival in mass extinction times. Yet survival or extinction in mass extinction episodes determines the large-scale shape of the tree of life. The massive culling of the synapsids (the so-called 'mammal-like reptiles') at the end of the Permian gave dinosaurs their chance. The death of the dinosaurs opened the door for the radiation of mammals.

(iii) In his *Wonderful Life*, Gould describes an extraordinary fauna from early in the history of animal life. The 'Burgess Shale' fauna is known in unexpected detail because a fortunate chance preserved soft tissues, not just the hard parts, of a large number of its members. Gould argues that this fauna demonstrates a completely unexpected pattern in the large-scale history of life. To make his point, he distinguishes between *diversity* and *disparity*. Life's diversity is the number of species in existence at that time. Gould accepts that life over the last few million years is probably more diverse than it has ever been. The disparity of life is measured by the number of basic organisations or body plans that exist at a time. The great richness of beetle fauna contributes hugely to life's diversity but not to its disparity. For beetles are

89

all built in the same general pattern, despite their variations in size, colour and sexual ornamentation. Having made this distinction between disparity and diversity, Gould makes three bold claims about the overall history of animal life. First, the disparity of animal life was at its peak early in its history. If the fossil record is to be believed, multi-celled animals evolved about 570 million years ago. By the time of the Burgess Shale fauna, a little before 500 million years ago, disparity was at its peak, and it has since shrunk. Second, very little disparity has been generated since the Cambrian. Those lineages that have survived have shown profound evolutionary conservatism. Contemporary disparity largely consists in the surviving elements of that initial burst of evolutionary experimentation. Third, Gould argues that survival has been 'contingent'; if we replayed the tape of history from the earliest Cambrian, with small alterations in the initial conditions, we might well have a very different set of survivors.

(iv) In *The Spread of Excellence*, Gould argues that evolutionary trends are not the scaled-up consequences of competitive interactions among organisms. One much repeated example of an evolutionary trend is the evolution of horses. Horse evolution saw a switch, the story runs, from browsing on trees and bushes to grazing on grasses. This change in lifestyle drove changes in morphology. Horses became larger, with longer, higher-crowned teeth, and at the same time lost their toes as their feet turned into hooves. If this were the right view of horse evolution, the example would be a triumph of extrapolationism. The evolutionary pattern in the horse lineage would be the aggregation, the summary, of a multitude of interactions in particular populations, most of which had the same upshot. But Gould reinterprets this history. It is not a result of the competitive success of grazers with those characteristics over browsing horses and less well-designed grazing horses. Rather, Gould argues that this trend is really a change in spread of variation within the horse lineage. The horse lineage used to be species rich, with a wide range of horse lifestyles and sizes. But only a very few species survived, and those few happen to be largish horses. The average horse is larger now only because almost all horse species went extinct, and the few survivors happened to be somewhat atypical.

Gould thinks there is a similar story to be told about a much more dramatic feature of life's history: its increase in complexity over time. Once there were bacteria, and nothing more. Then there was bacteria and protists: single-celled organisms still, but larger and more complexly structured than bacteria. Now there are whales and redwoods; coral reefs and sharks; honey bees and their hives; leaf-cutter ants and their gardens; and us. There is a trend, it seems, to increasing complexity over time. Gould agrees that there is a sense in which this trend is real. But, as with horse evolution, this is a misleading way to see history. Bacteria are still with us in immense diversity and in vast number. The change, he argues, is best seen as an increase over time in variation in complexity. The distance from the least to the most complex living creature has increased. The real phenomenon to be explained is this increase in variation rather than an upward trend in average complexity. There is, Gould argues, no such trend.

In considering Gould's case against extrapolationism, two issues will loom large. Are the patterns in life's history that he claims to detect real? And do these patterns really show the existence of evolutionary mechanisms other than those operating at the scale of local populations?

8   Punctuated Equilibrium

In 1972, Gould and Eldredge argued that the fossil records of most species show no significant change from the time they are first identified to the time they disappear. The fossil record is incomplete. Many species are known from only a few strata, though the record of marine invertebrates – and Eldredge's own special group is the *trilobites* – tends to be somewhat more complete. Moreover, with a few spectacular and famous exceptions, only the hard parts of animals fossilise, such as shells, bones and teeth. So some changes would be undetectable.

Gaps in the fossil record make evolutionary change seem jerkier than it really was, for intermediates that did exist are not found as fossils. Moreover, appearance and disappearance as we work our way through the strata at some site may well be migrations in and out as conditions change, rather than evolution being skittish. Even so, Gould and Eldredge argued that the appearance of stability was not a mere effect of the gappiness and imperfection of the fossil record. In most cases,

Figure 3: This diagram is from James Valentine's *On The Origin of Phyla* (p. 190), and it illustrates the limits of the fossil record. The top panel shows the organisms present in a sea-floor community. The bottom panel shows the organisms whose presence we would detect if all we had were fossils of hard parts. The difference is very striking, especially when we recall that sea-floor communities are the ones best preserved in the fossil record. So this illustrates the best case, not the worst case.

this appearance of evolutionary stasis reflects reality. Most species come into existence relatively rapidly, having acquired their distinctive characteristics, and do not significantly change thereafter. They do not vary more over time than they do at a time.

This hypothesis has been misunderstood in two important ways. In some early discussions of the idea, the contrast between geological and ecological time was blurred. Hence Gould and Eldredge were interpreted as making a very radical claim: species originate more or less overnight, in a single step, with all their new structures present. But that was a misreading. Occasionally plant species do arise in this way by hybridisation between parents of different species. But it is certainly unusual for animal species to originate in a single generation. Gould and Eldredge agree that new structures are almost always assembled over a number of generations, rather than all at once by a macromutation. Speciation – the sundering of a single lineage into two – takes generations. So by 'rapidly', they mean rapidly *by geologists' standards*. Species often exist for a few million years, and the resolution of the fossil record is coarse. In many circumstances, a speciation process that took 50,000 years would seem instantaneous. Yet that is only 2.5 per cent of the 2 million years, say, that the species existed. So a species that took 50,000 years to evolve its distinctive phenotype and which then persisted without significant change, would certainly conform to the punctuated equilibrium pattern. It would look as if it had appeared suddenly, and then not changed.

In recent work, they have clarified a second misunderstanding. In claiming that species typically undergo no further evolutionary change once speciation is complete, they are not claiming that there is no change at all between one generation and the next. Lineages do change. But the change between generations does not *accumulate*. Instead, over time, the species wobbles around its phenotypic mean. Jonathan Weiner's *The Beak of the Finch* describes this very process. In wet years, there is selection for slender beaks that enable finches to eat small soft seeds. In dry years, there is selection for more robust beaks. These are suited for cracking the larger harder seeds available in droughts. Wet years are interleaved with dry ones, so there is no long-term directional selection. The mean size and shape of the finch beak wobbles to and fro. If this fluctuating environment persists over the long term, finch species will be in stasis, as Gould and Eldredge define it. There will be no long-term shift in finch phenotypes.

How frequently do species stay in stasis over their lifespans? This issue is still open, in part because the species we find as fossils are a very biased sample of the species that have lived. Obviously, we are much less likely to find fossil evidence of soft-bodied organisms, especially small soft-bodied organisms (for very small animals will not leave 'trace fossils' either, i.e. fossils of trails, tracks, burrows and other effects on the physical environment). We are also much less likely to find the remains of specialist, narrowly distributed organisms than of generalist, broadly distributed ones: there are ecological as well as morphological biases in preservation. Even so, the consensus seems to be shifting Gould's way: the punctuated equilibrium pattern is common, perhaps even predominant. So let's grant that stasis is common. Why suppose this is bad news for extrapolationist orthodoxy?

Gould and Eldredge agree that new structures are created by cumulative selection over many generations. No departure from orthodoxy here. Moreover, the example of the Galapagos finches shows that we can explain stasis by extrapolation from processes we can observe in local populations. If we aggregated the data from many Galapagos seasons, the result would be a 'wobble' around an average finch beak. There are other local processes that generate stasis, too. For organisms can track their preferred habitats as the environment changes, rather than staying put and adapting *in situ*. The fossil record of the Pleistocene shows that the geographic distribution of many animals shifted in response to climate changes, and we can see similar changes in range on human timescales. Many areas of Australia have been converted from woodland to grassland. As a result, the range of many grassland species has expanded, whereas that of other species has contracted. There are far more red kangaroos in Australia now than there were when Europeans arrived in 1788. So once punctuated equilibrium is stripped of its radical misreadings, how does it collide with gene-selectionist neo-Darwinian orthodoxy?

Gould and Eldredge are right in thinking that the processes we can observe in local populations do not tell the whole story. That picture needs to be supplemented. The problem is not stasis but speciation. How can events in a local population generate a new species? This question is at the heart of Eldredge's *Reinventing Darwin*, which revisits the debate with Dawkins. In answering this question, Eldredge develops an idea of Douglas Futuyama. His short answer is that they usually do not. Local populations change, as the example of Australian rabbits and myxomatosis shows. But changes in a local population are usually too fragile to make a new species. Adaptation to local circumstances as, say, an impala population adapts to a particularly dry region, usually depends on gene complexes rather than single genes. Such complexes in local populations are vulnerable to being swamped, either by migration or by one population fusing with another. Local populations are short-lived and their boundaries are permeable, and so the clock of local evolutionary change is always in danger of being set back to zero. The facts which make stasis easy to explain make speciation hard to explain.

Nonetheless, speciation is obviously possible. New species do come into existence. There are different ideas on how this puzzle is to be solved. In his most recent work on this issue, Eldredge thinks of speciation as a threshold phenomenon. Small changes in the environment change the distribution of species: such changes wipe out some populations and amalgamate others. Only larger disturbances impose both isolation and novel selective regimes rigorously enough to make new species. The idea that species-making depends on external disturbances rather than co-evolutionary interactions seems one-sided to me, but any solution to the speciation problem will take us beyond events in local populations observable on human timescales. For example, Elisabeth Vrba (another of Gould's co-authors) argues that occasional climate changes are responsible for a *turnover pulse*. These changes rob some species of their entire habitat, and those species go extinct. But other species will be fragmented. Some of their populations may change their character. Instead of being semi-isolated, or briefly isolated, they will be fully isolated, and for long periods. Most such isolated fragments will sooner or later go extinct. But a few will become new species, for changes within them will accumulate rather than being washed away by fusion within the larger population.

These models probably have only partial validity. But it is likely that whatever explains the occasional transformation of a population into a species will rely on large-scale but rare climatic, biological, geographic or geological events; events which isolate populations until local change is entrenched (the so-called 'Milankovich cycles' is one such isolating mechanism). This is an exception to extrapolationism. We cannot understand speciation just by studying evolutionary change in a local population. The prospects of a local population becoming an incipient species depend on the ways that species is fractured into populations and on the relations between those populations. So speciation is not just the accumulation of events in a local population; it depends as well on the way that population is embedded into a large whole. There is a break here with a strong version of extrapolationism, but it is not a radical break. Dawkins could, should and probably would accept it; in *The Ancestor's Tale*, he has an inclusive view of speciation mechanisms. After all, Ernst Mayr, one of the architects of contemporary Darwinism, has long defended a view of speciation along these lines. Gould somewhat overstates the adherence of orthodoxy to strict extrapolationism.

Punctuated equilibrium is more important than the rather ungenerous treatment that Dawkins gave it in *The Blind Watchmaker*. For he interpreted it as an idea about the rate of change in local populations. I see it as a thesis about how, and under what circumstances, local changes become speciation events. If these circumstances are unusual, and if, as Vrba and Eldredge argue, there is a common pattern to them, then the theory of speciation is an ingredient we need to add to the gene selectionist's tool kit. Ernst Mayr taught evolutionary biologists that speciation occurs only when a fragment of an ancestral species becomes isolated (most usually, by geographical separation) from the rest of the population and diverges from that ancestor. Vrba and Eldredge add to Mayr's original insight a theory of when such fragmentation is most likely to take place, and when the fragments are likely to survive rather than go extinct. Punctuated equilibrium is, then, an important idea.

9 Mass Extinction

Extinction is normal. Species go extinct as a result of local ecological interactions. The Stephens Island wren lived only on (surprise!) Stephens Island (in New Zealand), and is now extinct as a result of wren/cat interactions. Other species disappear as a result of competitive displacement. Still others are just unlucky; they are unfortunate enough to be on a volcano that erupts or in a lake that dries out. These examples pose no problem for the view that the evolutionary history of species and species lineages is just an aggregate of local ecological processes of the kind we can and have observed. For we do observe hunting cats, erupting volcanoes and drying lakes. However, Gould argues that many extinct lineages did not die the death of a thousand microevolutionary cuts. They have gone out not with a myriad of whimpers but as part of a big bang. Major lineages in the tree of life typically go extinct in periods of mass extinction; periods which change the rules of the evolutionary game.

Earth history is divided into Eons, Eras, Periods

and Epochs largely by the contrasts in their species composition. Thus the divide between the Permian and the Triassic, about 260 million years ago, is one of the most profound divides of earth history. It marks the close of the Palaeozoic era and the beginning of the Mesozoic. Douglas Erwin's *The Great Paleozoic Crisis* begins with a 'before and after' snapshot of typical marine invertebrate communities across this boundary. The shift is dramatic. The Permian community is dominated by filter-feeding animals attached firmly to a base. Most animals are immobile. There are exceptions: mobile animals like fish, cephalopods (squid, octopus and their relatives), snails and bivalves (clams, oysters, mussels and the like) are part of the community. But they are present only in small numbers. In contrast, Mesozoic communities are dominated by animals which move under their own steam. Most of the fixed-in-place animals of the Permian, together with the coral species which built the reefs on which the former established themselves, have gone. The biological world had changed massively.

These divisions in earth's history seem to imply the existence of some great killing episode. For if there were not a particular event, *a specific killer*, why would deaths in, say, the snail lineage (the gastropods) be correlated with those of the starfish and sea urchins (the echinoderms) or with terrestrial reptiles? The very organisation of geological history seems to presuppose a real difference between successive biotas, a

Figure 4: The advent of the Mesozoic era saw an explosion in the number and diversity of mobile taxa in marine communities. The first picture shows a Palaeozoic reef community, the second a late-Mesozoic bottom community. (Source: Drawn by Mary Parrish, in D.H. Erwin, *The Great Paleozoic Crisis*, New York: Columbia University Press, 1993.)

difference with a relatively unified cause. If so, massive extinctions must be of profound importance, reshaping the tree of life. In these episodes, major *clades* – founding species and all their descendants – disappear. The trilobites disappeared in the end-Permian extinction, and the ammonites are never found beyond the Cretaceous/Tertiary boundary. But even those clades that survive are often profoundly altered. Erwin points out that though the echinoderm and snail lineages survived the Palaeozoic crisis, only a few species in each survived. Their diversity was profoundly reduced, and that reduction in diversity marked the rest of the history of those groups. One lineage of dinosaurs, the birds, did survive the great killing at the Cretaceous/Tertiary. But while the birds are a diverse and species-rich group (there are about 10,000 living species), they have not recaptured the biological territory vacated by their extinct relatives. A 'Modern fauna' can be snail rich, but still unlike the Palaeozoic snail fauna. Some surviving lineages never come back at all; in David Jablonski's haunting phrase, they are 'dead clades walking'. In short, in Erwin's view, the Palaeozoic crisis reshaped the whole history of life, and the same seems likely to be true of other mass extinction events that remove major players from world ecosystems.

As always, matters are not so simple. The trilobites disappeared for ever at the end of the Permian. But the diversity of the trilobite lineage had already shrunk profoundly before that disaster. Their extinction may have even pre-dated the close of the Permian. John Archibald, in his *Dinosaur Extinction and the End of an Era*, defends similar views of the dinosaurs, arguing that their range and biological diversity had already shrunk before the end of the Mesozoic. If this line of thought is right, and mass extinctions merely speed up a process already underway, then they would not make much difference in the very long run. If mass extinction's effects are selective, and the less well-adapted species are those most likely to go extinct, major crises – periods when many species die – might simply accentuate evolutionary trends already in process. In a series of books (most recently, *Nature: An Economic History*), the distinguished palaeo-biologist Geerat Vermeij has argued that the basic history of life is an arms race between predators and prey. On that model, extinction will mostly be driven by the very processes we can observe locally: superior competitors will squeeze out less effective ones. David Raup calls these *fair game* models of mass extinction.

In evaluating the fair game model, it is important to discover the nature and duration of mass extinction episodes. It is virtually certain that the earth was hit by a massive extraterrestrial object at roughly the time of the end of the Mesozoic mass extinction. Many think this impact causes the extinctions. If that is right, if the changes that cause extinctions are genuinely sudden – sudden in an ecological time frame, as catastrophic climate changes caused by an impact would be – then selective accounts of extinction become very implausible. For the survival of a species would depend on the suite of biological characters it happened to have at the moment of change; whether these are fortunate or unfortunate. But if mass extinction episodes take place over millions of years, then extinction can be sensitive to the evolutionary response of the lineage. The more structured, gradual and smeared out over time a mass extinction is, the more likely it is that the distinction between global event-driven mass extinction and background extinction is a difference in degree, not kind. There is a very recent suggestion that some mass extinction events are not the results of external crises at all, but reflect a slowdown in the rate at which new species are born. Species go extinct, but for the usual routine reasons: climate change, competition, local bad luck. If, but only if, mass extinction episodes are fast and discontinuous with the events that surround them, then mass extinctions will have distinctive evolutionary effects. They will change, perhaps profoundly, the history of life. That is *exactly* Gould's hypothesis: he thinks that they are 'more frequent, more rapid, more intense, and more different in their effects' (*The Structure of Evolutionary Theory*, p. 1313) than we have supposed.

This issue has been debated most vigorously in connection with the death of the dinosaurs. No one now seriously doubts that there was a meteor impact at the Cretaceous/Tertiary boundary. But there is still a lot of debate on its significance. After all, if that is all that happened, why did crocodiles, turtles and even frogs go through relatively unscathed? One line of argument suggests that the ammonites, together with dinosaurs (other than birds), pterosaurs, pliosaurs and other marine reptiles had all shrunk significantly in diversity and range before the impact. Perhaps some of these groups were already extinct, and so talk of the 'Cretaceous/Tertiary extinction' compacts the process in time. It treats events that had been happening over millions of years as if they happened in a geological instant. In the case of dinosaurs, perhaps the meteorite only administered the final blow to a group on the way out. Others argue that the great clades that failed to make it to the Tertiary were in good shape before unforeseeable catastrophe overtook their world.

The view that mass extinctions essentially accentuate processes already under way cannot be the whole truth. The changes mass extinctions impose are too vast. That is particularly true of the catastrophe that struck life at the end of the Permian. Probably more than 90 per cent of animal species then alive went extinct, and it may well have been rapid ecologically, not just geologically, beginning with massive volcanic eruptions in Siberia. So, at least, Michael Benton argues in his *When Life Nearly Died*. Extinction on this scale must have caused fundamental reorganisations of life. If so, we cannot understand the overall history of life by projecting, onto the largest scale, processes we see operating in local populations. Mass extinctions are not just local bad news scaled up.

Moreover, Gould, drawing on the work of David Raup, argues that there is a distinctive evolutionary regime in operation in periods of mass extinction. These are not casinos ruled by chance alone. There are principles which would enable us to pick winners and losers. The game has rules. But they are different rules from those of normal times. The magnitude of the upheaval at the Permian/Triassic boundary, and the pace of the upheaval at the Cretaceous/Tertiary (if the impact was important), make it unlikely that the game was fair. Adaptation, we recall, is adaptation to a specific environment. Scramble the environment – drop a polar bear in the desert – and even a species superbly adapted to its previous environment will be in deep trouble. So, as Raup puts it, extinction was probably 'wanton'. Species survival is not random, but the properties on which survival depends are not adaptations to the danger mass extinction threatens. If a meteor impact caused a nuclear winter, then the ability to lie dormant would have improved your chances. But dormancy is not an adaptation to the danger of meteor impacts.

The ability to lie dormant is a characteristic of individual organisms. But many important characteristics relevant to survival or extinction would have been properties of species themselves. Species with broad geographic ranges, species with broad habitat tolerances, species whose lifecycle does not tie them too closely to a particular type of community all would have had a better chance of making it. At least, it is plausible to suggest this, though testing these suggestions empirically turns out to be very difficult. In any case, Gould argues that survival and extinction through periods of mass extinction involve some form of species selection. If so, mass extinctions are doubly important. They restructure the history of life, and they do so in part through a sieve, a filter, not visible in local populations in local communities. For filters in local communities are sensitive to the properties of organisms, not species. Thus Gould links his views on the role of mass extinction in evolution to another of his ideas, one we have seen before. According to him, selection does not just act on individual organisms, it acts on collectives built out of organisms. One such collective, in Gould's view, are species themselves. He suspects (but not claims to prove) that patterns of survival in mass extinction are profoundly influenced by species-level properties. What looks like dumb luck, if we concentrate just on the features of individual organisms, would be predictable if we were to scale up our visual capacities, and see the species as a whole.

The idea that species themselves are selected is interesting and provocative. But it is also very problematic. There are, as Gould himself concedes, no well-worked-out case studies: no examples of patterns clearly predicted given the species selection hypothesis and only that hypothesis. Thus there are no good examples of the idea working. But it is not even clear that the idea could work. It is true that if punctuated equilibrium describes the predominant pattern of a species life history, species do indeed have identifiable births and deaths. There are also, as we have just seen, species-level, macro-properties, and these may well make a difference to a species' prospects for surviving extinction dangers. But those macro-properties may not be inherited by daughter species. The ancestral finch species that found its way to the Galapagos and became Darwin's finches had a broad geographic range and tolerated a wide range of conditions (otherwise it would never have arrived and survived). But many of its daughter species have highly specialised lifestyles and narrow geographic ranges. The individual finches still look pretty much like their ancestor, but the macro-properties of the descendant species are not similar to those of their ancestor.

If this is true and typical, there cannot be meaningful selection on species themselves. Selection and evolution requires a population of individuals in competition, who breed true, and who have differential fitness. We have just seen that when we are thinking about macro-properties, species-level properties, perhaps species do not breed true. Moreover, fitness differences between species may have no evolutionary consequences. For fitness and success are not quite the same. The success and failure of individual organisms is not always the result of their fitness. Some mortality is the result of unhappy accident rather than ill-design, and some animals are fecund by luck, despite their design flaws. In the terminology of biology, actual fitness is not always identical to expected fitness. It is only on average that the fitter design will do better. The chance of its doing so increases with the number of trials, just as the chance that a fair coin will land heads up about 50 per cent of the time increases with the number of throws. Three heads in a row is not so unlikely; 30 and someone is cheating. It follows that in a small population, luck matters more. There will be fewer versions of each design, and hence a greater chance that the best design will die from a not so improbable run of bad luck. The number of species in a clade is small by comparison to the number of organisms in a species, and so unlucky extinctions and lucky survivals are likely to be both frequent and important. Even small fitness differences usually matter in the world of organisms, for often population sizes are large, and the different variants within the species face selection for many generations. The different varieties within a population get trialled repeatedly and frequently. That is not true of populations of species, even if they do breed true.

In short, Gould's case for the importance of mass extinction depends on the view that there is a qualitative difference between mass extinction and background extinction, and that major groups have disappeared that would otherwise have survived. While hard to prove, that claim is very plausible. Gould adds to this claim an extra and optional element: the idea that species-level properties in part determine survival, so mass extinction regimes are species selection regimes. Once more, that is a plausible but much more controversial conjecture. It awaits confirmation.

10 Life in the Cambrian

Conventional wisdom emphasises the gradualness of evolutionary change. New organs – circulatory systems, nerve nets, limbs and tentacles, perceptual organs – are put together bit by bit over countless generations. So are new ways of organising tissues and organs into functional animals. In this respect, Dawkins is a true son of orthodoxy. He cannot remind us too often that the power of selection to build our exquisite and intricate biota depends on its slow and incremental operation. Every living creature is a triumph over chance. No random process, no hurricane blowing through a junkyard, could ever assemble anything as wildly improbable as a flea or a weevil. Each organic design is a victory over the improbable, and each is won by insensible degree. Mount Improbable is ascended by the smoothest and gentlest of tracks.

This standard story seems to run slap bang into a nasty fact. About 543 million years ago, at the base of the Cambrian, the Ediacaran fauna apparently disappeared. There were living animals, for enigmatic small shelly fossils have been found from these times. Moreover, there are fossilised tracks and then burrows, showing that something moved over and then through those ancient seafloors. But as far as we can tell from fossil evidence, the diversity of animal life at the beginning of the Cambrian Period was very limited. By the middle of the Cambrian, about 520 million years ago, animal life was rich and diverse. That diversity is spectacularly shown in the Chengjiang fossils of China. These are as spectacular as the Burgess Shale fauna, and significantly older. Thus the fossil record seems to show that most of the major animal groups appeared simultaneously. In the 'Cambrian explosion', we find segmented worms, velvet worms, starfish and their allies, molluscs (bivalves, snails, squid and their relatives), sponges, brachiopods and other shelled animals appearing all at once, with their basic organisation, organ systems and sensory mechanisms already operational. We do not find crude prototypes of, say, starfish or trilobites. Moreover, we do not find the common ancestors of these groups. Multi-celled animals are probably a monophyletic clade: a single ancestral species that gave rise to all, and only, the animals. Modern groups arose from this common ancestor. So, for example, there must have been within this group a single lineage of animals which were the ancestors of the so-called 'deuterostomes'. We are deuterostomes (as are all the other chordates), as are starfish and their relatives and an obscure group of worms (acorn worms) and their relatives. The deuterostomes share a common and distinctive developmental pattern in their early embryo, and a set of genetic similarities. So it does seem clear that there was a single lineage which has this diverse and unlikely group as their descendants. But no likely candidate has been found in the fossil record.

This explosive evolutionary radiation of the Cambrian seems to be unique. Plants seem to have arisen somewhat more gradually. Flowering plants evolved well after the gymnosperms, and gymnosperms were preceded by earlier plant lineages, some now wholly extinct. Nor was there a similar radiation when animals invaded the land. Dry land – and even fairly moist land – provided empty ecological space for the first animals which could adapt to living out of water. But the colonisation of the land saw no proliferation of wholly new ways of making an animal. Arthropods remain recognisable arthropods; and snails, snails. Vertebrates retain the basic vertebrate organisation. Worms, too, stay worm-like, even as representatives of all these groups acquired the special adaptations needed for life in the dry. So perhaps evolution worked under different rules back then. As we shall see, Gould is inclined to think so.

One alternative is to argue that the 'explosive Cambrian radiation' of animal life is an illusion generated by the failure of earlier Precambrian fossils to survive to our times. Darwin and many of his successors supposed that the emergence of the animals only *seemed* explosive. They thought the first appear-

Figure 5: Trilobites (i) are arthropods, starfish (ii) are echinoderms and velvet worms (iii) are onychophora. Arthropods, echinoderms and onychophora are living lineages and hence the discovery of representatives of early members of these lineages in the Cambrian was no surprise, even though onychophora are no longer found in the ocean. Break open rotting wood in New Zealand, however, and you have a decent chance of finding one. Hallucigenia (iv), one of Gould's initial examples of extraordinary disparity, turned out to be a mistakenly assembled velvet worm. Like it, the trilobite and the starfish are 'regular' animals. Trilobites are no longer with us, but

iii

iv

they are abundant in and well known from the fossil record. Moreover, they are obvious relatives of those living marine arthropods, the crabs, crays and other crustaceans. Starfish (when we stop to think of them) really are weird in their own right, with their five-sided symmetry and their strange system of circulation and locomotion. But they are comfortably familiar. Other fossils from the Cambrian, though, are the remains of much stranger animals. (Sources: i, ii, iii: C. Tudge, *The Variety of Life*, Oxford University Press, 2000; iv: Drawn by Marianne Collins, in S.J. Gould, *Wonderful Life*, London: Hutchinson Radius, 1990.)

ance of multi-celled animals in the fossil record was preceded by a long history of hidden evolution. This remains a live option. There are fossil embryos of animals from China dating to about 570 million years ago. Moreover, there are many lineages of animals with no fossil record at all: James Valentine reminds us that there are twelve phyla which have no fossil record. If early animals were small and soft-bodied, they may have lived for many a million years without leaving traces that we can detect (of course, there would still be a puzzle about why they remained so small and unobtrusive for so long). Moreover, there are, of course, Precambrian animal fossils, the Edia-carans we met in Chapter 1. These are named for the place they were first found and recognised for the strange beasts they seem to have been. For shortly after the Second World War, fossils of Precambrian animals were discovered in Ediacara, South Australia. Subsequently it has been shown that this Ediacaran fauna was present worldwide. So there was animal life before the Cambrian. However, the relationship between the Ediacaran fauna and that of the Cambrian remains very unclear. In one view – a view Gould is inclined to support – the Ediacaran fauna were a failed experiment in the history of life; a branch of the tree of life that was wholly extinct before or at the Cambrian. The Ediacaran animals were not ancestors of the Cambrian animals and hence their existence does not extend the time frame of animal evolution into the Precambrian.

Ediacaran fossils are not the only reason for suspecting the existence of a hidden history that pre-dates the Cambrian explosion. The last few decades have seen the development of molecular methods of estimating the time at which two lineages diverged. When, say, the velvet worm lineage diverged from that of the arthropods, each lineage inherited its genetic material from that last common ancestor. Once the lineages began to evolve independently, differences started to appear in their genetic material. If we can measure the extent of that difference, and calibrate the rate at which DNA sequences diverge from one another, we can estimate the time the last common ancestor lived. The idea is to compare a stretch of DNA in the velvet worm lineage with an equivalent stretch in an arthropod, and measure the extent to which they have diverged from one another. If the rate of divergence can then be calibrated from lineages with rich, detailed fossil records, we can estimate the time the last common ancestor of the velvet worms and arthropods lived. Moreover, we can get *molecular clock* dates for those lineages that have no fossil record. This is how we know the zero-fossil phyla are old, too. This technique has many potential uncertainties. It depends on careful calibration of the rate of change of genes. It depends on careful choice of the genes used. For example, suppose we were to choose the genes that code for the neural networks of the two animals. If those genes had been subject to strong selection in, say, the arthropods, then that selection would have produced genetic change. Arthropod neurone-making genes would have become quite different from velvet worm neurone-making genes. If selection on arthropods had been more intense than on the lineage we used to calibrate the molecular clock, we would overestimate the depth in time of the divergence of the two lineages.

The strength of selection, and hence the rate of change it causes, can obviously vary over time and between lineages. So these uncertainties can be reduced by choosing genes which are not being changed by selection. Many genes are silent, coding for no protein at all. Some of these silent genes have no effect on an organism's phenotype. Such mutations are 'neutral', and are not visible to selection. The same is true of changes to a gene which make no difference to the protein it builds. Such changes are possible because the gene/protein code is redundant. The protein-building machinery reads DNA sequences in sets of three bases, and quite often a change in the third position makes no difference to the protein that is made. The general presumption with molecular clocks is that the rate of neutral changes will not vary much between lineages or over time; this is based on the idea that the mechanisms which copy genes and correct errors are ancient and highly conserved, and hence have been inherited in similar versions by all the lineages whose evolution we are considering. So we can increase our confidence in molecular clocks by choosing silent genes or genes which code for highly conserved features of organisms; for example, for very basic metabolic functions common to all animals. Neither silent nor conserved genes change as a result of selection. A second way is to use a number of DNA sequences not just one. If two, three or more different clocks all give similar dates of divergence, we can be more confident that they are approximately right.

At first, the molecular clock dates for the evolution of the main animal lineages were very surprising indeed. The molecular clocks indicated that the common ancestors of the main animal lineages lived hundreds of millions of years before the Cambrian. The last common ancestor of, say, the trilobite lineage and the flatworm lineage may have lived over 800 million years ago. Despite the uncertainties of molecular clocks, it is increasingly clear that the animal phyla first recorded in the Cambrian does indeed have a hidden evolutionary history. The last common ancestor of the animal phyla lived a very long time before the Cambrian. More recently, though, the molecular dates and the fossil dates have been less brutally discordant. Those who have revised the very old dates argue that they depended on using the fossil record of the vertebrates for their calibration of the clock, and that vertebrates and their genes evolve slowly, and hence the dates using those calibrations were all too old. The whole issue remains very controversial, and the last word has certainly not been said. Overall, though, it is probably fair to say that the most

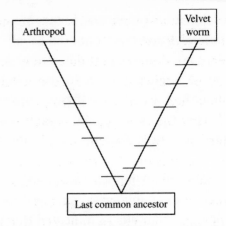

Figure 6: A history in which the distinctive morphologies of velvet worm and arthropod lineages are built step by step from the time of initial divergence.

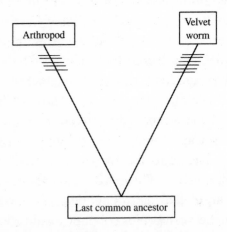

Figure 7: A history in which the distinctive morphologies of velvet worm and arthropod lineages develop in a quick burst, long after the initial divergence. (Arthropods and velvet worms are relatives, and both are segmented, but they are probably not one another's closest relatives: their last common ancestor would also have been an ancestor to other animal lineages.)

recent developments have reduced the mismatch between molecular and fossil data.

However, even the youngest dates from molecular clocks place the origins of the deepest branches in the tree of animal life – where the sponges and jellyfish branch off from the other early animals – over 600 million years ago, and so quite deep in the Precambrian. Gould accepts this, but rightly points out that it remains possible that the Cambrian explosion really was explosive. For we need to distinguish between the origin of a lineage and the evolution of its distinctive bodily organisation and ecological role. The molecular clock (when all goes well) dates the origins. The fossils date geographical spread and morphology. The divergence of two lineages is one thing; the acquisition of their distinctive organisation and equipment is another. In Figures 6 and 7, the horizontal bars indicate evolutionary changes in the lineage. So Figure 6 depicts a history in which those distinctive morphologies are built step by step from the time of initial divergence. Figure 7, on the other hand, depicts an evolutionary history in which the velvet worm and arthropod lineages diverged in the deep past, but developed their distinctive morphologies in a rapid evolutionary burst long after their initial divergence. Gould points out that molecular clock data cannot decide between these two possibilities.

Moreover, Gould argues that the fossil record supports the model in which the lineage splits much earlier than the distinctive morphologies evolve.

For that explains why we find no Precambrian proto-arthropod fossils. In short, the 'hidden history' hypothesis remains an open option, but so does Gould's guess that the Cambrian explosion was genuinely explosive rather than an illusion generated by incomplete preservation. Evolution at the base of the Cambrian may really have been especially fast. Perhaps competition was weak in an empty world. Perhaps the developmental programs of early animals were more open – less constrained – and so more variation was generated.

Gould's ideas about the rapidity of the early Cambrian radiation form the least controversial of his claims about this era of life. For he thinks recent evidence about the Cambrian fauna overthrows our standard conception of life's history, a picture that sees life as becoming more diverse and better adapted over time. As he sees it, a remarkable discovery made in the early years of the 20th century overthrows this conception. The discovery was that of the Burgess Shale fauna; a discovery remarkable because the Burgess Shale preserved not just bits of shell and bone, but soft structures as well. So it yielded a record of creatures without hard parts, creatures whose existence would otherwise never have been suspected. And it reveals much more than we would usually know about those creatures with hard bits. More recently, two other remarkable Cambrian sites have been discovered: Chengjiang in China and the Sirius Passet formation in Greenland. Both are significantly older than the Burgess fossils; the Chinese site is about 522 million years old; and the Greenland formation is dated at about 518 million years before the present, compared to the Burgess Shale which was laid down about 505 million years ago. These older sites show a rich and diverse fauna and hence are important for the claim that the radiation was truly explosive. Almost 40 million years passed from the beginning of the Cambrian to the Burgess Shale fauna. On anyone's views, a lot of evolution can happen in 40 million years.

The distinguished American palaeontologist, Charles Walcott, discovered the Burgess Shale fossils, but interpreted these creatures as simpler versions of well-known types of animal. During the 1980s, this interpretation was radically revised. The revision suggested that many of the Burgess animals were radically unlike anything living. Some were recognisable arthropods. But they were not members of any of the four great arthropod groups: spiders and their allies; trilobites; crabs, lobsters and the like; or insects and insect-like forms. So, the idea ran, the Cambrian saw not just the invention of the arthropods – segmented, jointed-limbed, exoskeleton-covered animals – but of far more kinds of arthropods than have ever been seen since.

As we saw in Chapter 7, in developing this idea, Gould distinguished between the *diversity* of life and its *disparity*. The diversity of life is the number of species extant at that time, and no one doubts that the diversity of life has increased since the Cambrian.

Disparity is not counted by species numbers. Disparity measures the morphological and physiological differentiation between species. The discovery in recent years that New Zealand has not one but two species of *tuatara* was the discovery of extra diversity in New Zealand. (The tuatara is a lizard-like reptile; the sole living survivor of the sister group to snakes and lizards.) But it was not the discovery of extra disparity. For the two species are so alike that the existence of separate species was only suspected when molecular techniques showed different populations on different islands were genetically quite dissimilar, even though structurally they are almost identical. However, the 19th-century discovery in New Zealand of the tuatara, like that of the echidna and platypus in Australia, was the discovery of significant additional vertebrate disparity. For the *monotremes* are unlike other mammals not just in laying eggs. They are structurally unlike other mammals, too, in having a single opening that serves for both reproduction and excretion. They also have very unusual adaptations: they electrolocate their invertebrate prey, and the male platypus is the sole mammal armed with venom and a method for injecting it.

Armed with the distinction between disparity and diversity, we can now explore Gould's more radical claims about the history of animal life. As Gould sees it, though diversity has increased, disparity has radically shrunk since the Cambrian. Animal life was at its most disparate at the peak of that explosion. The arthropods are by far the largest clade of animals, and there were more fundamentally different kinds of arthropod alive in the Cambrian than the world has seen since. The same is true at an even grander scale. The major subdivisions of animal life are phyla. Each phylum is a distinctive way of building an animal. The molluscs, for example, jointly form a single phylum. Among the Burgess Shale fauna, Gould says, we find many animals that are members of no surviving phylum. And they are as different from the modern phyla – from the velvet worms, arthropods, molluscs, vertebrates, flatworms, starfish, bivalves and the others – as each of those phyla are from one another. To put the point succinctly: there were many phyla alive then that are no longer around. With a single exception, all the living phyla that have reasonable fossil records are found in the Cambrian.

In brief, the Cambrian phylum count was larger, maybe much larger, than the contemporary count. No new phyla have appeared, and many have gone. That count, in turn, is a reasonable measure of disparity. So Cambrian disparity was considerably greater than current disparity. The history of animal life is not a history of gradually increasing differentiation. It is a history of exuberant initial proliferation followed by much loss; perhaps sudden loss. If all of this is right, it raises some very fundamental questions about life's history. Why was the Cambrian so rich in disparity, and why was the disparity generated so rapidly, and then lost?

i

ii

Figure 8: If the trilobites and echinoderms seem reasonably familiar, these animals do not. Perhaps they are *stem-group* arthropods, but *Opabinia* (i) and *Anomalocaris* (ii) are strikingly, obviously different from anything living in today's oceans. Animals such as these make Gould's views about Cambrian disparity seem very plausible. But is this just the shock of the old, or is there some objective sense in which they are very different from living animals and trilobites are not?

Gould rather doubts that selection has much to do either with the early burst of disparity, the conservative nature of evolution since the Cambrian, or with the roster of loss and survival. If no or few phyla have evolved since the Cambrian, and if counts of phyla measure the disparity of life reasonably well, the history of animal life since the Cambrian has been strikingly conservative. No new plans have been invented; no old ones have been massively modified. If Gould is right about this basic pattern of history, he is surely right in thinking we are faced by a mystery. If early animal life was highly disparate, and if little new disparity has subsequently evolved, we need to know why. For evolutionary change in general has not ground to a halt in the last 500 million years. That period has seen the evolution of all the adaptive apparatus needed for life on, under and above the ground. Many adaptive complexes, not least human intelligence, have been invented in that period. So why have new phyla – major new ways of organising animal bodies – not been invented too?

Dawkins and, even more, his former student Mark Ridley think the basic claim about the pattern of history is not right. They challenge Gould's views in two ways. Ridley, in particular, is sceptical of the distinction on which the whole picture rests. He is a committed cladist. Cladists have a very distinctive conception of the aim of biological classification. They think that a biological classification is an evolutionary genealogy. The *cladogram* in Figure 9 makes no

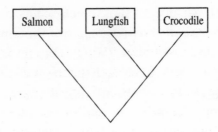

Figure 9: Lungfish are more closely related to crocodiles than they are to salmon, yet systematicists once classified lungfish and salmon together on the basis of morphological similarities.

claim at all about the morphological, physiological or behavioural similarities of salmon, lungfish and crocodiles. Rather, it claims that lungfish and crocodiles share a more recent common ancestor than either share with the salmon. The purpose of biological systematics is to discover and represent genealogical relationships between species. And that is its only purpose. Cladists think that the only groups we should recognise and name – genera, families, orders, classes and phyla – are *monophyletic*. A monophyletic group consists of *all* and *only* the descendants of a single founding species. Mammals are monophyletic: a single ancestral species gave rise to all and only the mammals. The group of animals we call reptiles, by contrast, is not monophyletic. There is no species that had all and only the reptiles as its descendants. Birds and mammals also descended from the common ancestor of all the reptiles. So cladists do not think of the reptiles as a real group.

The cladists' reasons for focusing on genealogical

relations strike at the very idea of disparity. They do not think we can objectively measure similarity. Similarity and dissimilarity are not objective features of the living world. Our judgements of similarity and difference reflect the biases of human perceptions and interests, not the objective features of the world. We are visual organisms; sight is our primary sense. We are therefore very struck by differences in visual appearance. One of Gould's weird wonders of the Burgess Shale, *Opabinia*, has a cluster of five eyes. It certainly looks genuinely odd. But imagine we were equipped with noses like a bloodhound, and that scent was as rich a source of information for us as vision. Perhaps we would then think of the moths, say, as varying wildly from one another, by virtue of differences in the feather-like organs they use to detect the pheromones they use to attract mates. From the perspectives of moths, other moths have exuberantly and wildly different noses. In short, organisms differ from one another in their morphology and physiology in endless ways. Some of those differences are more salient to us, more striking and surprising, than others. But according to the cladists, that is a fact about us, and the ways we interact with the world. *It is not a fact about the history of life*. A sentient electric eel, given the same data, would reconstruct the genealogy of life – who is related to whom – in the same way we do. For the pattern of descent is an objective fact of history, though one difficult to discover. But would they make the same judgements of disparity? Cladists doubt it.

Like many evolutionary biologists, Gould uses spatial metaphors to explore and explain his ideas, and he recruits a spatial metaphor to explain disparity. 'Morphospace' is the space of animal designs. In Chapter 6 we saw a cut-down region of morphospace, a three-dimensional space which represents all possible shells. This three-dimensional space enabled us to represent shell disparity, and all the possible ways shells might be shaped. It turned out that real shells occupy only a small region of that space. To represent the disparity of real organisms, even those that live in shells, we need a morphospace with many more dimensions. Three dimensions do not, for example, capture the contrast between oysters and brachiopods. Both are shelled animals, but brachiopods have a very distinctive feeding apparatus (and are notoriously inedible). Three-dimensional shell morphospace tells us nothing about the organisation of the animal in the shell. One way of putting the cladists' point is to ask: how do we determine the dimensions of morphospace? How do we measure distance in any given dimension? We can measure an infinite number of features of any given animal. We can measure the number of leg hairs on a fruit fly. Is this a dimension of morphospace? What about the ratio of the facets of its compound eye to their leg hair number? If that sounds absurdly recondite, it is not. Insect taxonomists routinely tell one species from another by measuring the curves, crinkles, hooks and barbs on their sexual organs. For disparity to be an objective feature of the tree of life, there must be a principled way of answering questions such as these. There must be some principle that shows, for example, that variation in leg number among arthropods is a genuine aspect of disparity, whereas variation in nostril hair number in primates is not. Cladists doubt that any such principle is to be found. Gould accepts that this challenge is difficult to meet, but believes palaeobiology can and must develop ways of meeting it.

Moreover, cladists think they can explain why Gould's weird Burgess wonders seem so weird, even though they are not. When Gould first wrote about the Cambrian fauna, the cladists had not made much progress in building a family tree of the animals, showing who was most closely related to whom. In the last decade, using shared genetic characteristics rather than morphological ones, a lot of progress has been made in building this animal family tree. In the process, cladists think that they can explain why Gould thought these animals were so very unusual. Their crucial idea was to distinguish between the stem and the crown members of a lineage. I shall borrow an example from Andrew Knoll's wonderful *Life On a Young Planet* to explain the idea and its importance. Knoll's example focuses on the evolution of the arthropods, and their (possible) sister lineage, the nematode worms. If these are indeed sisters, there was a lineage that gave rise to the arthropods and nematodes, and to no other living organisms. Arthropods and nematodes do have some features in common: they share a moulting cuticle together with a number of genetic and developmental signatures. Despite those similarities, they are very different animals. Arthropods are segmented, with jointed appendages and an external skeleton. Nematodes are tiny worms, tapered at both ends. If these animals really did have a common ancestor, it could hardly have been much like the signature arthropods. But nor would it have had the radically simplified anatomy of these tiny worms.

So imagine the evolutionary history of the arthropod lineage leading from this last common ancestor to the crustaceans, trilobites, insects and spiders. The features that make us recognise these animals as arthropods – the segmentation, the jointed limbs and body, the exoskeleton – evolved. But all at once. Perhaps the order was: first the outer armour, then the segmented body, followed by the jointed arms, legs and feelers, thence jointed appendages. If this is even roughly right, there will be *stem group arthropods*: arthropods in the lineage that lie between the last common ancestor of the nematode–arthropod group and the first typical arthropod. That first typical arthropod – the first full-deal arthropod – and all its descendants are the *crown group arthropods*. Since the arthropod lineage began with animals which must have been quite unlike full-deal arthropods, we know there must have been aberrant arthropods. In the lineage that began with the nematode–arthropod split, there were species living before the evolution of full-deal arthropods. If we discovered fossils of those creatures, they would have to look weird to us: they would be somewhat like full-deal arthropods, but not wholly like them.

Nematodes, which are so very unlike arthropods, make this very vivid. But the sister lineage of the arthropods, whatever it turns out to be, must be morphologically very distinct from the arthropods. For the arthropods themselves are such a distinctive group of animals; nothing else is much like them. Hence the last common ancestor of the arthropods and whatever animals turn out to be their closest relative will not look much like a full-deal arthropod. In turn, it follows that early stem group arthropods must be very different from full-deal arthropods. Now stem arthropods gave rise to the crown group arthropods – the living arthropods plus those extinct arthropods which share the distinctive characteristics that unite the living arthropods. But crown arthropods are not the only animals which evolved from stem arthropods. There were side-branches as well. And while the lights eventually went out for all those side-branches (for otherwise we would not call them side-branches), in some cases these lineages lived and changed for millions of years before their extinction. *Opabinia* and the *Anomalocaris* are probably examples of such early experiments in arthropod evolution. So the distinction between stem group animals and crown group animals explains why early arthropods seem so different and so disparate, without supposing that there was anything unusual about Cambrian evolution or Cambrian disparity. Cambrian stem group arthropods are bound to have unique characteristics possessed by no living animals. And they will also have some of the features of full-deal arthropods, but not all of them. The fact that they are hard to classify using categories which have been invented for more recent animals is exactly what we should expect. The perception of extraordinary Cambrian disparity is a kind of illusion. We should expect these early animals to stretch our biological categories even if the evolutionary mechanisms that built them are exactly the same as those that built the many species of fruit fly on the Hawaiian archipelago.

Dawkins, like these cladists, thinks that Gould is suffering from an illusion, but for a somewhat different reason. He too thinks that Gould over-counts Cambrian disparity. To understand Dawkins' point, we need to make a short venture into arthropod taxonomy. The arthropod trunk of the tree of life is divided into four great branches – four classes – that, except for the extinct trilobites, have been evolving independently of one another for 500+ million years. These classes are the trilobites, the insect-like arthropods, the crustaceans, and the spider-like arthropods. Systematicists identify these animals on the basis of the pattern of bodily segments, and the number and pattern of limbs and feelers on those segments. Living arthropods exemplify three basic patterns. Crustaceans, for example, have a basic division into the head and the trunk. The trunk varies a good deal, but the head is always divided into five segments, each with a pair of branched limbs. Of these five pairs, two are antennae, two are maxillae and one is mandibles. Systematicists are interested in these features because they are good markers of the genealogical relations between arthropods. The evolutionary genealogist needs traits that show some, but not too much, evolutionary plasticity relative to the lifespan of the group of interest. The possession of an exoskeleton or the habit of laying eggs are important traits. But they are uninformative: they are universal among the arthropods. They are too *conservative*. Other traits, like the number of segments which form the body of the animal, are too *variable* for keeping track of the basic pattern of arthropod relationships.

It turns out that the basic pattern of body segmentation, the pattern of limb development (in particular, whether limbs develop as single or as branched structures) and the number of segments forming the head are features of arthropod history that have been evolutionarily conserved over periods of 500 million years. They are conservative enough, but not too conservative. Once an arthropod lineage has evolved, say, with the crustacean pattern of limbs on its head, animals later in that lineage do not lose that pattern. And animals outside that lineage do not evolve the trait independently. So it acts as a membership badge for that branch of the arthropod family. In contrast, over long periods the possession of eyes is not a good marker of evolutionary relationships. For eyes are lost as well as gained, and they have often evolved independently. Now Gould takes traits – segmentation and appendage patterns – that are genuinely important as markers of genealogical connection among the arthropods, and treats these as measures of arthropod disparity. He identifies extraordinary levels of arthropod disparity in the Burgess Shale fauna on the grounds that this fauna exemplifies segmentation and appendage patterns not found in the four great branches of arthropod life today. So what? As we have just seen, these traits are not chosen for their intrinsic importance. They are chosen because they have become conservative over periods of 500 million years. There may be no special significance in crustaceans having five pairs of appendages on their heads. That may be just a minor but frozen historical accident. That pattern still indicates relatedness. The point is familiar from human genealogy. An unusual surname is of no intrinsic significance, but it can still indicate a family connection.

So even if, contra Ridley's suspicions, disparity is a genuine property of the tree of life, there is no reason to suppose that it is measured by traits which are suitable for keeping track of relatedness over long periods of time. The distinction between disparity and diversity is very plausible. Some of the Burgess animals really do look strange and wonderful. Even so, it is fair to say the Ridley/Dawkins challenge is yet to be met. We lack a good account of the nature of disparity, and we lack objective measures of it. Without that, the existence of Gould's puzzling pattern remains conjectural.

11   The Evolutionary Escalator

Three and a half billion years ago, the most complex and sophisticated life forms were cynobacteria. Cynobacteria are not just single-celled organisms. They are single-celled organisms lacking a nucleus, mitochondria, chloroplasts and a raft of other internal structures. Many eukaryotic single-celled organisms now exist. These are organisms with parts very nearly as complex as a bacterium (probably because they once were bacteria). Still more obviously, multi-celled plants and animals have evolved. These are not just huge bags of cells. They are differentiated assemblies. Animals consist of an array of cell types which are organised into tissues, organs, organ systems and so on. This is an astounding evolutionary achievement. As an animal or plant grows from a single fertilised cell, it does not just get larger through cell division. As the cells in an animal divide, at some stage they have to begin turning into neural cells, muscle fibres, blood cells, sex cells and the tissues of an array of specialised organs. Eyes, for example, need cells with photoreceptors. The specialised cells have to be assembled into larger structures – tissues and organs – and connected appropriately with others. We showed in Chapter 5 that cooperation problems are hard. Despite that, these differentiating cells are solving such a cooperation problem. But they are also solving a coordination problem. A cell's form depends on where it is in the body and on what other cells are doing. Tissue and organ formation is interdependent. As all this growth and reorganisation takes place, the embryo has to stay functional enough to stay alive. In many species, it even has to fend for itself. So the life of every complex organism depends on the solution of a complex cooperation and coordination problem.

It is easy to make it seem as if development could not happen. Yet millions of different developmental packages have evolved in the last billion years or so, and the result is an extraordinary array of multi-celled organisms. These are now capable of surviving and reproducing in an array of habitats from the highest mountains to the very deepest parts of the oceans. These habitats were invaded in stages by both plants and animals. Even after animals established themselves on land, it took time for desiccation-resistant eggs to evolve, to free reproduction from dependence on water. Plant evolution, too, showed a similarly incremental infiltration of terrestrial habitats. Fairly primitive plants had established themselves by the beginning of the Devonian at about 410 million years ago. By the close of the Devonian, at about 360 million years ago, plants with bark, leaves, wood and seeds (but not yet flowers) had evolved. Creeping towards modernity took about 50 million years.

In the light of all this, surely it is obvious that the history of life on earth shows a progressive increase both in complexity and adaptedness. Gould does not quite reject this view outright. But he thinks it is a very misleading way to think about the history of life. His reasoning connects the direction of life's history with a major theme of his work: the importance of non-selective explanations of broad patterns in the history of life. Gould offers a reinterpretation of evolutionary trends. This reinterpretation includes this largest-scale trend of all: the tendency of life to become increasingly complex over time.

Let's begin with horses. As early as the 19th century, when Thomas Huxley championed Darwin's ideas, the history of horses was one of the paradigms of evolutionary change. In response to the opportunity provided by the evolution of grass, and the establish-ment of grasslands, horses became prairie animals rather than forest animals: their size, their feet and their teeth changed in response to their new world. So at least the standard story has it. But Gould thinks that this trend in horse evolution is a mirage. What has really happened in the horse lineage is a loss of diversity. There has not been a directional trend in horse evolution. Rather, there has been massive extinction in that lineage and the paltry surviving remnant happens to be grazers. The appearance of a trend is generated by a reduction in the heterogeneity of that lineage. We need only to imagine a slightly different possible world in which the new world horses (*Nannippus*) lived and the old world species (*Equus*) did not, and the alleged trend would disappear from the rocks (*The Structure of Evolutionary Theory*, pp. 905–08). A trend which is hostage to one switch between life and death is no trend at all.

In discussing complexity, Gould tells a similar story at the very largest scale. What we think of as a progressive increase in complexity is a change in the difference between the least and the most complex organism. It is a change in the spread of complexity. Life starts off as simple as life can be. Physics and chemistry impose constraints that define the least complex possible form of life. Bacteria are probably close to that limit, so life starts at the minimum level of complexity. Since even now nearly everything that is alive is a bacterium, for the most part life has stayed that way. But evolution will from time to time build a lineage that becomes more complex over time. No global evolutionary mechanisms prevent more com-plex organisms evolving from less complex ones. But none make it more likely. The complexity of the most complex life form alive tends to drift up over the generations, just because the point of origin of life is close to the physical lower bound. Relative to bacteria, there are never many of these complex creatures. But the difference between the simplest and the most complex organism tends to become greater over time.

If life originates close to the point of minimum complexity, wholly undirected mechanisms will increase this range. Given the existence of a minimum limit, and an origin close to that limit, mechanisms that are *blind to complexity* suffice for an *upward drift* in average complexity. But bacteria continue to dominate the living world. So it is misleading, at best, to think that evolution is characterised by a trend towards increased complexity.

Gould's picture is illustrated by the two snapshots of life in Figures 10 and 11. The first is of life soon after its origin. There is little variation in complexity. Everything alive is close to the lower bound – the left wall – of complexity. The second is a snapshot of life a few billion years or so later. The mode has not changed. Most living things are still close to the left wall. But the curve has spread to the right, not to the left. For there

Figure 10: A snapshot of life soon after its origin. Here there is little variation in the complexity of organisms.

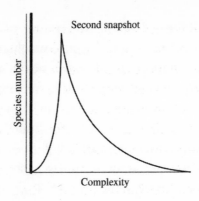

Figure 11: A second snapshot of life a few billion years later. Here the variation in the complexity of organisms has increased.

is a wall imposed by the laws of the physical sciences to the left, but none to the right.

From Dawkins' perspective, Gould's picture gets something important right, but it is not the whole truth. It leaves out an important aspect of genuine progress in the history of life, and it understates the dynamism of evolution. Evolution changes its own rules, in ways this picture leaves out. In general, Dawkins thinks that the course of evolutionary history is more predictable than does Gould. In *Wonderful Life*, Gould introduced a famous metaphor: 'replaying the tape' of life's history. Gould used that metaphor to ask an important but difficult question. Suppose we began the history of life again (from its very beginning, or from some crucial event), but in somewhat altered conditions. Would the replayed tape resemble in broad outlines life's actual history? Would animals evolve again? Would land be invaded? Would we see something like trees again? According to Gould, we have no reason to suppose that such an alternative history would at all resemble our actual history, even if it began from a similar starting position. As he sees it, mass extinction events are so important, in part, because they make history utterly unpredictable. Simon Conway Morris, in *Life's Solution*, responded, arguing that convergent evolution is such a ubiquitous feature of evolution that the broad outline of evolution is highly predictable. Evolutionary pathways are constrained by both opportunity and possibility. There are not very many ways of building working organisms, and so we can predict that evolution will move along this small set of pathways. Many of the most distinctive features of living systems have evolved more than once. Some of them (like eyes) have evolved many times.

In *The Ancestor's Tale*, Dawkins accepts the general ideas behind Conway Morris' argument. A replay of the tape of life from similar beginnings would result in a somewhat similar history, with many of the most striking adaptations on our world – multi-celled organisms, social life and the division of labour, well-constructed detection systems, well-constructed locomotion systems – evolving again. In particular, we will expect to see progressive evolution again. For Dawkins, in contrast to Gould, thinks that in an important though limited sense, our evolutionary history is one of progressive change. However, Dawkins has a somewhat different take on the relationship between progress and complexity than does Gould. Gould is interested in complexity only because he regards complexity as a measurable indicator of progress. Dawkins does not – he thinks the link between progress and complexity is anthropocentric. He thinks evolution is progressive because, over time, life is becoming *better adapted*.

Life does not become better adapted in every respect. The physical conditions of existence change: the climate becomes hotter, dryer, wetter, windier, colder or more seasonal. Organisms become adapted to local conditions. Then local conditions change, and organisms must move or readapt. There is no reason to suppose that there is any arrow of overall improvement here. The same is not true, Dawkins thinks, of the relationship between an organism and its enemies. Predators and their prey; parasites and their hosts are locked into a permanent arms race. An improvement in a predator's efficiency makes the prey's environment worse, and in a predictable and continuing way. It imposes permanent rather than fluctuating selection in favour of a response. The same is true of the effect of prey on their predators. An improvement in camouflage, in vigilance, in escape behaviour makes the prey's environment worse, in a predictable and permanent way, likewise imposing an ongoing selection pressure. This will be true however often selection on predators and their prey throws up improvements. So lineages are locked together in arms races generating progressive change. Both predator and prey become absolutely more efficient in hunting and avoiding hunters, though their relative success with respect to one another may not change at all over time.

So as Dawkins sees it, progress is real though partial and intermittent. It is partial because it is generated only when selective regimes are both directional and stable: selecting for the same kind of phenotype change over long periods, as in arms races. It is intermittent because every arms race will ultimately be disrupted by large-scale environmental changes. Arms races between dinosaur lineages, obviously, came to an end when their world was blown away, setting up the conditions for new races between herbivores and plants; birds, mammals and their parasites, and so on. Thus in *The Ancestor's Tale*, Dawkins compares progressive evolution to the teeth of a saw: incremental improvement 'plunged abruptly at the end of the Cretaceous, when the last of the dinosaurs abruptly gave way to the mammals' new and spectacular climb of progressive improvement' (p. 618). Arms races between lineages are cut short by mass extinction events, but while they are in progress each lineage is objectively improving.

Gould overstates his case. There is more to the history of life's complexity than a gradual increase in variance. In 1995, John Maynard Smith and Eörs Szathmáry published *The Major Transitions in Evolution*. Their view of life's history involves a series of major transitions and hence an inherent directionality.

Each of these transitions opens the door to the evolution of more complex living creatures. Life's possibilities change over time, with these possibility-expanding innovations. Some of these transitions involve the change from naked replicators to the first organisms. But they also think of the invention of eukaryotes; cellular differentiation and the invention of plants, animals and fungi; and the origin of social life as major transitions. Dawkins, too, has defended a similar series of major transitions, though not in as much detail, in discussing the evolution of evolvability at the end of *The Ancestor's Tale*. There Dawkins discusses a series of 'watershed events', events which make new forms possible. He nominates the evolution of sex, the invention of multi-celled life together with a life cycle that takes large organisms through a single-celled reproduction stage, and the invention of a modular mode of the development and construction of bodies as major watersheds in evolvability. Segmentation, for Dawkins, is a special case of modularity: of building a creature out of relatively discrete chunks. For once a chunk has been invented by evolution, it can be modified or redeployed without stuffing up the rest of the organisms.

Gould, too, is interested in evolvability, but he does not discuss evolvability itself evolving over time (*The Structure of Evolutionary Theory*, pp. 1272–7). A crucial difference between Gould, on the one hand, and Maynard Smith, Szathmáry and Dawkins, on the other, is a difference in how they picture the spread of complexity. Gould thinks of complexity as having a lower bound but no upper bound, and these features of complexity are fixed by biochemistry, not by the course of evolutionary history. Over time, given that life originates near the lower bound of complexity, the spread increases as it trickles away from the limit imposed by the minimum possible complexity. Maynard Smith and Szathmáry do not regard the walls as fixed over time. Until the foundations of eukaryotic life were gradually assembled, there was also an *upper bound* on complexity. That bound was set by the intrinsic limits on the size and structural complexity of prokaryotes. For much of its history, for perhaps 2 billion years, bacterial evolution was confined between these two limits. Similarly, after the evolution of the eukaryote, there was a shift of the upper limit, but only a relatively small one. The invention of the multi-celled organism required a complex series of evolutionary innovations. Until these came into existence, there was an upper bound to complexity set by the limits on a single eukaryotic cell. Maynard Smith and Szathmáry argue that social existence, too, has evolutionary preconditions. Until these are met, a wall remains to the right. So, where Gould sees unchanging boundaries set by physics and chemistry, Maynard Smith, Szathmáry and Dawkins see evolution as transforming these boundaries irreversibly. The eukaryotic cell, sexual reproduction and cellular differentiation all change the nature of evolutionary possibility. These possibilities have changed over time in a direction that increases the maximum attainable complexity. In short, over time the rules of evolution change.

Evolvability has changed. For what can evolve depends on the developmental mechanisms that determine the variation available to selection. These have changed over time. For example, once cellular differentiation had been invented, new variation was made available to selection. These changes opened up new possibilities, especially the possibility of more complex life forms. In *Full House*, Gould insists that this age, and every age, is the age of bacteria. Bacteria are the world's most numerous organisms. They have the most disparate metabolic pathways. They may well add up to most of the world's biomass. All this is true and important; Dawkins himself says much the same in chapter 39 of *The Ancestor's Tale*. But it is not the whole truth. We live in an age in which many biological structures are now possible that were once not possible. That too is true, and important.

Part IV
The State of Play

12   A Candle in the Dark?

Dawkins and his allies really do have a different conception of evolution from that embraced by Eldredge, Lewontin and other collaborators of Gould. But those differences do not explain the undercurrent of hostility this debate has generated; hostility that surfaced dramatically in the *New York Review of Books* exchange. No doubt some of that hostility has a banal psychological explanation. People in general do not much enjoy being told they are wrong, especially in public. A somewhat prickly response is no great surprise. But I doubt whether this is the whole story. Dawkins and Gould mostly argue about issues internal to evolutionary theory. But they have very different attitudes to science itself.

Dawkins is an old-fashioned science worshipper (here I line up with him, not Gould). Like all scientists, he accepts the fundamental Popperian point that scientific theory is always provisional, always open to revision in the light of new evidence and new ideas. And he accepts, of course, that in the short run human error and human prejudice can block our recognition of important evidence and good ideas. But Dawkins is wholly untouched by the postmodern climate of current intellectual life. For him, science is not just one knowledge system among many. It is not a socially constructed reflection of the dominant ideology of our times. To the contrary: though occasionally fallible, the natural sciences are our one great engine for producing objective knowledge about the world. In many cases, we can be confident that received scientific opinion is right, or very nearly right. And that knowledge is liberating. In short, for Dawkins science is not just a light in the dark. It is by far our best, and perhaps our only, light.

Dawkins does accept a fact–value distinction, as the first essay in *A Devil's Chaplain* makes clear. Science cannot tell us what to accept and reject. But he does not think of values as a special kind of fact that can be studied non-scientifically, by the humanities or social sciences (though there are of course facts about the values we embrace, facts which anthropologists can and do study). Least of all does he think religion has any special authority on values. Gould's take on the status of science is much more ambiguous. For one thing, he thinks some important questions are outside science's scope. He defends this idea in his recent work on the relationship between science and religion. On this issue, Dawkins' views are simple. He is an atheist. Theisms of all varieties are just bad ideas about how the world works, and science can prove that those ideas are bad. What is worse, as he sees it, these bad ideas have mostly had socially unfortunate consequences. Gould, by contrast, seems to think theism is irrelevant to religion. He interprets religion as a system of moral belief. Its essential feature is that it makes moral claims about how we ought to live. In Gould's view, science is irrelevant to moral claims. Science and religion are concerned with independent domains.

Gould's views on religion are doubly strange. First, it seems extraordinary to overlook the innumerable claims about the history of the world and how it works that are made by the various religions. The claim that the world was created by a being with intentions and expectations seems to be a factual claim about the world, not a moral claim about what we should do. Furthermore, those factual claims are often supposed to be the basis of a religion's moral injunctions. So they are not minor details of religious belief systems we can reasonably ignore.

Second, Gould seems to commit himself to a very strange conception of ethics. Does he think that there are genuine ethical truths? Is there genuine moral knowledge? Recent thinking on ethics has gone two ways on this question. Perhaps the main contemporary line of thought is to argue that moral claims are expressions of the speaker's attitude or intentions towards some act or individual. According to this 'expressivist' view, when, for instance, I call someone a scumbag, I do not describe a particular moral property of that person. Rather, I express my distaste for that person and their doings. The main alternative is some version of 'naturalism'. From this viewpoint, moral claims *are* factual claims. They are based on facts, though typically very complex facts, about human welfare. Gould seems to commit himself to denying both these options. If 'expressivism' is right, there is no independent domain of moral knowledge to which religion contributes. Our moral utterances are not designed to describe objective features of the world, but are instead vehicles for expressing our attitudes and emotions. I suspect Dawkins is some kind of expressivist. Alternatively, if naturalism is right, science is central to morality. For it discovers the conditions under which we prosper. The appeal to religion has largely dropped out of the picture. For one thing, religions really do seem to make claims about the world, and ones that cannot be rationally sustained. For another, even if these claims were true, they do not seem to give us any *moral reason* for action. This point was made in classical Greek civilisation, and it can be condensed into a single question: 'Is torturing babies bad because God forbids it, or does God forbid it because it is bad?' Give the first answer, and you are committed to the bizarre view that it would be right, not just prudent, to torture should God command it. Give the second, and you concede the irrelevance of religion to moral truth.

So Gould thinks that there are important domains of human understanding in which science plays no role. Moreover, he is much more sceptical about the role of science, even in its 'proper' domain. Even so, he certainly rejects extreme versions of postmodern relativism. It is an objective fact of evolutionary history, and one that we know, that dinosaurs evolved by the Triassic, dominated terrestrial ecosystems during the Triassic, Jurassic and Cretaceous, and (with the exception of the birds) went extinct at the Cretaceous/Tertiary boundary, probably as a result of a large meteor striking the earth. There is no sense in which this is just a Western creation myth, a reflection of the dominant ideology of these times, or just an element of the current palaeobiological paradigm. It really happened that way, and we really do know that it did. So *to some extent* Gould shares with Dawkins the view that science delivers objective knowledge about the world as it is.

Scientific belief is sensitive to objective evidence. It is more than a mere reflection of the culture and values of its times. But Gould argues that science is very deeply influenced by the cultural and social matrix in which it develops. Many of his *Natural History* columns illustrate both the influence of social context on science, and also its ultimate sensitivity to evidence. Indeed, this became an increasingly pervasive feature of these columns. They began as reflections on natural history; they ended as reflections on the history of natural history. This sensitivity of science to its cultural location need not distort it. Sometimes the influence is beneficial, providing useful metaphors and models. Darwin's debt to 19th-century political economy is the most famous such example. In *Time's Arrow, Time's Cycle*, Gould locates the development of our conception of deep history in its cultural and intellectual context without any suggestion that that cultural context perverted the development of geology. But these influences are often constraining. In *Wonderful Life*, Gould argued that the Burgess Shale fauna were misunderstood because they were interpreted through the ideology of their discoverer. Walcott expected to see evidence of slow but progressive evolution and he shoehorned the fossils into his preconceived picture. Especially when the scientific issues are directly relevant to social and political concerns, all too often these socio-cultural interests have led to bad science, pseudo-science, racist and sexist science. *The Mismeasure of Man* is Gould's most famous essay on these themes. In it he is concerned to show how a particular ideological context led to a warped and distorted appreciation of the evidence on human difference.

So, one sharp contrast between Dawkins and Gould is on the application of science in general, and evolutionary biology in particular, to our species. This is surely a source of much of the underlying tension in this debate. Dawkins' most systematic writings on human evolution, a little paradoxically, explore the differences between human evolution and that of most other organisms. For in human evolution, *memes* – ideas and skills – are important replicators. Ideas are copied generation by generation, just as genes are.

Tunes, football allegiances, ethnic prejudices and skills are copied from one human to another. We humans are vehicles of the memes we carry, not just the genes we carry. Notoriously, Dawkins has exploited this machinery to give a vivid and provocative analysis of religious belief. The crucial idea is to see us as ecosystems as well as organisms; we travel through life with many passengers. Some are physical, and some are informational. Some benefit us; some do not. Stable cooperation depends on a shared fate, when the success of one depends on the success of all. In a submarine under attack, there is no temptation to cheat, for everyone will survive or no one will. This logic applies to genes and organisms (as we saw in Chapter 5). Each of us is home to myriads of bacteria and viruses. And many of us harbour some assortment of parasitic worms, protists, fungi and lice. Many of these need not share out fate. They can spread to our friends and colleagues. Thus it can pay a virus to sacrifice me to enhance its own travel to new hosts. That is why viruses are sometimes virulent. Dawkins argues that our ideas are often like pathogens or parasites. They are replicators, but their fate is not tied to one another or to the fate of our nuclear genes. For we spread our thoughts to friends, colleagues and acquaintances, and we get information from such sources too. This openness has great benefits: we can piggyback on the discoveries of others without taking the risks that make those discoveries possible. But we also pay a cost: a meme that is costly to its host can still spread. Those with truly destructive genes cannot pass them on, because a person passes their genes only to their children, and those with killer genes have no children. Those with truly self-destructive ideas have no children either, but they do still have a chance of passing on their lethal views.

In 'Viruses of the Mind' (in *A Devil's Chaplain*), Dawkins sees religion through this lens. Not all religions are evangelical; Judaism, for example, is mostly a family affair. People of this persuasion have almost always inherited their views from their parents. But some religions are, and followers of these spread their views, like the flu, to their friends, colleagues, neighbours and children. More zealous, evangelical religions are more likely to be destructive to their bearer, because they depend less on parent–child contagion. Dawkins' own picture of cultural evolution may not work. There are doubts about whether ideas are copied reliably enough and accurately enough for his specific picture to work. But his general approach has recently become very popular indeed. In *Not By Genes Alone*, Peter Richerson and Robert Boyd show that it is impossible to understand human lives and culture without an evolutionary approach. Eytan Avital and Eva Jablonka, in *Animal Traditions*, extend these models from the human to the animal world, showing that cultural evolution has played an important role in the evolution of other animals. So though Dawkins approaches human behaviour using different tools to those of standard sociobiologists and evolutionary psychologists, he is fully committed to the idea that we can understand ourselves only in an evolutionary framework.

The contrast with Gould is deep. Of course, Gould accepts that we are an evolved species. But everything Gould does not like in contemporary evolutionary thinking comes together in human sociobiology and its descendant, evolutionary psychology. The result has been a twenty-year campaign of savage polemic against evolutionary theories of human behaviour. Gould *hates* sociobiology. It is true that some evolutionary psychology does seem simple-minded. For example, Randy Thornhill's work on rape is unconvincing. He argues that in some circumstances sexually excluded men can improve their fitness by acts of rape, but he makes no attempt to take into account the fitness costs of sexual violence, and neglects obvious and serious problems for the idea that a tendency to rape is an adaptation. It is tempting to believe that *The Natural History of Rape* is a deliberate provocation.

Many contemporary evolutionary psychologists have taken on board the need for caution in testing adaptationist hypotheses. That is even more clearly true of those whose background is in biological anthropology rather than cognitive psychology: Blaffer Hrdy's *Mother Nature* certainly does not assume that every characteristic is an adaptation. However, even those more cautious defenders of sociobiology, and its intellectual descendants, downplay the aspects of evolutionary process central to Gould's thought. They tend not to emphasise the importance of development and history in imposing constraints on adaptation, the problems in translating microevolutionary change into species-level change, the role of contingency and mass extinction in reshaping evolving lineages, or the importance of palaeobiology to evolutionary biology. Sociobiology, even at its most disciplined, reflects a different angle on evolution to that exemplified by Gould. This must play some role in his hostility. But most of all, I suspect, Gould thinks these ideas are dangerous and ill-motivated as well as wrong. They smack of hubris, of science moving beyond its proper domain, and incautiously at that. Dawkins does not concur. For him, knowledge of the evolutionary underpinnings of human behaviour is potentially liberating rather than dangerous. For example, he takes Axelrod's work on cooperation (work he discusses in the second edition of *The Selfish Gene*) to be a reason for optimism about our condition. Likewise, an evolutionary perspective on human life might even help us escape the poisoned chalice of religion.

13　Stumps Summary

Let's remind ourselves of the fundamental contrasts between Dawkins' views and those of Gould, and then summarise the state of the debate.

Dawkins argues that:

(i) Selection fundamentally acts on lineages of replicators. Most replicators are genes; chunks of DNA. But not quite all: in animals capable of social learning, some replicators are ideas or skills. And the earliest replicators were certainly not genes.

(ii) Genes typically compete by forming alliances which build vehicles. In such cases, genes succeed or fail through their distinctive, repeatable influences on these vehicles. A gene which generation by generation enhances the sensory acuity, metabolic efficiency or sexual attractiveness of a vehicle will be replicated more frequently than its rivals.

(iii) Some genes have other replication strategies. Outlaws enhance their prospects at the expense of the adaptive design of the vehicle. Extended phenotype genes advantageously engineer the physical, biological or social environment of the vehicle they are in.

(iv) Dawkins is not committed by the logic of his position to the view that vehicles are always individual organisms – groups might be vehicles. But the existence of cooperation between animals is no reason to think that groups of animals, rather than the individual animals in the groups, are vehicles. In the right circumstances, it pays individual animals to cooperate.

(v) The central problem evolutionary biology must explain is the existence of complex adaptation. So, natural selection has a special status within evolutionary biology, for complex adaptation can only be explained by natural selection.

(vi) Complex adaptation evolves gradually. Very occasionally, a copying error in a replicator may result in a large yet survivable phenotype change. But almost all new adaptive complexes and new ways of building an organism evolve gradually, through a series of small steps.

(vii) From the perspective of evolutionary biology, humans are an unusual species. For they are vehicles of memes as well as genes. Nonetheless, the basic intellectual tools of evolutionary biology – especially those explaining cooperation, reciprocation and sociality – also apply to human evolution.

(viii) Extrapolationism is a sound working theory. Most evolutionary patterns are accumulations over vast stretches of time of microevolutionary events. Phyla – the great lineages of animal life – began as ordinary speciation events and they grew the same way. But not quite all evolutionary patterns fit the extrapolationist view. Possibility-expanding changes, for example, might result in some form of lineage-level selection. Lineages in which possibility-expanding changes evolve (segmentation, sexual reproduction, a desiccation-resistant egg) are likely to be more species-rich and disparate as a result.

As we have seen, Gould's picture is quite different. As he sees matters:

(i) Selection usually acts on organisms in a local population. But, in theory and in practice, selection acts at many levels. Groups of organisms can form populations of groups, with some groups doing better than others. Within a lineage of species, some may have characteristics that make them less likely to go extinct, or more likely to give rise to new species. It is even possible to have selection acting on individual genes within an organism, though this is the exception not the rule. An evolutionary change at one level often affects the options for future change at other levels.

(ii) Many characteristics of individual organisms are not explained by selection. Furthermore, there are important patterns in the large-scale history of life that have no selective explanation. Selection is important, and evolutionary biologists must understand its operation. But it is just one of many factors explaining microevolutionary events and macroevolutionary patterns. Moreover, complex adaptation is only one of the phenomena evolutionary biology must explain.

(iii) Extrapolationism is not a good theory. Large-scale patterns in the history of life – in particular, those tied to mass extinction episodes – cannot be understood by extrapolation from events we can measure in local populations.

(iv) Evolutionary biology needs a theory of variation: a theory that explains the variation available to an evolving lineage. The supply of variation forecloses some potential changes and makes others more likely.

(v) Humans are evolved animals. But attempts to explain human social behaviour that make use of the techniques of evolutionary biology have largely been failures, vitiated by a one-sided understanding of evolutionary biology. They have often been biologically naive.

These debates are still alive and developing, so no final adjudication is possible yet. But we can say something about how the argument has developed. The idea that gene-selectionist views of evolution are tacitly dependent on reductionism and genetic determinism is a mistake. Dawkins and the other gene selectionists do not think that nothing happens in evolution but changes in gene frequency. They do not deny the immense significance of the evolution of the organism. Rather, they see the evolution of organisms as the evolution of vehicles of selection; of 'survival machines', as Dawkins has called them. These machines interact with other survival machines, and with the inanimate environment, in a way that ensures the replication of the genes whose vehicles they are. But the construction of organisms is not the only strategy available to genes to enhance their prospects for replication. One way the gene-selectionist view of evolution contrasts with others is in its emphasis on these other strategies; on outlaws and extended pheno-type genes. Extended phenotype genes are common and important, for the parasitic style of life is very common – there are millions of parasite species – and probably all parasite gene pools include genes whose adaptive effects are on host organisms. The outlaw count is unknown, but it is growing all the time. Outlaws may well turn out to be more common than we had supposed.

Gene selectionism is not genetic determinism. No gene selectionist thinks that there is typically a simple relationship between carrying a particular gene and having a particular phenotype. There *are* such genes but these are the exception not the rule. A given gene – say, the human sickle-cell haemoglobin gene – exercises phenotypic power that promotes its own replication prospects only in a given context. Change the context from one in which that gene is paired with a normal haemoglobin gene to one in which it is paired with another copy of itself, and you change the resulting phenotype. Gene-selectionist ideas are certainly compatible with the context dependence of gene action. But they do suppose that there is some reasonably regular relationship between the presence of a particular gene in the genotype of an organism, and an aspect of that organism's phenotype. In talking about lineages of magpie aggression genes, rabbit disease-resistance genes and host manipulation genes, gene selectionists assume that genes in those lineages affect their vehicles in fairly similar ways. So, while gene selectionists are not genetic determinists, they are making a bet on developmental biology. When relativised to reoccurring features of context, gene action will turn out to be fairly systematic. There is no reason to suppose this hunch is false, but it is not known to be true.

Developmental biology is salient to this debate in a second important aspect: the role of selection in evolution. Gould is betting that when the facts of developmental biology are in, it will turn out that the evolutionary possibilities of most lineages are highly constrained. The envelope of potential variation available in, say, a crustacean lineage is restricted to fairly minor alterations of that lineage's current organisation. Krill, for example, carry their gills external to their carapace, giving them a characteristic feathery appearance. Gould is betting that characters of this kind are 'frozen' into the lineage. They are developmentally entrenched. That is, these basic organisational features are connected in development to most aspects of the organism's phenotype, and that makes them hard to change. A mutation which affected the location of gills on the krill will affect many aspects of krill phenotype. Some of these effects would certainly be deleterious. For most changes in a functional system make that system less effective, not more effective. Since variation in these frozen-in features is unlikely, selection is not likely to be important in explaining their persistence. Gould thinks that 'frozen accidents' play an important role in explaining evolutionary patterns revealed by the fossil record. Dawkins' bets are different. Over time, selection can alter the range of evolutionary possibilities in a lineage. So he thinks both that selection has a larger range of variation with which to work, and that when patterns do persist over long periods – hundreds of millions of years – selection will have played a stabilising role.

The integration of evolution and development is the hottest of hot topics in contemporary evolutionary theory, and this issue is still most certainly open. The contrasting bets on developmental biology are still undecided, but are subjects of active research. My best current guess is that developmental biology probably does generate biases in the variation that is available to selection, and hence that evolutionary trajectories will often depend both on selection and these biases in supply. The working assumption of the modern synthesis was that the supply of variation was not biased. On this older view, it was supposed that variation in a trait is distributed densely and without bias around its current mean. If the average mass of a possum is two kilos, we will find as many lighter possums as heavier ones; we will find more possums just a bit heavier or lighter than average than we will find possums that are a lot heavier or lighter than average. However, this picture of the supply of variation was based on a particular view of the relationship between genes and bodies: the idea that each characteristic depends on a large number of genes, each of which had a small effect. If that were true, mutations, too, would usually have small effects, ones that shifted the trait just a bit away from its average. But mutation is not restricted to 'point mutations': mutation which might substitute one amino acid for another. Mutation can result in movement, duplication, inversion and deletion of DNA sequences, and hence can result in changes to gene regulation and to shifts in reading frames as well as changes to the amino acids that are transcribed from the new DNA. So 'middle-size' mutations, mutations which make a significant difference to traits, are likely to be important in many evolutionary changes. These may well introduce biases in the supply of variation. In brief, Gould's hunch that developmental biology plays a crucial role in explaining evolutionary patterns will probably be vindicated.

It is harder to see how to resolve some of Gould's other claims about the large-scale history of life. Despite the great plausibility of the distinction between disparity and diversity, we are not close to constructing a good account of disparity and its measurement. Even granted that distinction, it is hard to see how to test Gould's idea that the large-scale history of life is contingent; his idea that if we 'replayed the tape' with minor variations in the starting setting, the outcome would be dramatically different. Obviously we cannot perform that experiment. And there are no natural experiments at a large enough scale. As we saw in Chapter 11, Conway Morris, in *The Crucible of Creation* and *Life's Solution*, argued that convergent evolution shows that history cannot be as unpredictable as Gould supposes. In convergence, two independent lineages come to resemble one another when both face similar environmental pressures. Old and New World vultures, for example, are not closely related birds, but they are very much alike in terms of appearance and behaviour. But there are problems with this line of thought. First, most examples of convergence are not independent evolutionary experiments. For they concern lineages with an enormous amount of shared history, and hence shared developmental potential. This is true of the standard example of convergent streamlining in marine reptiles, sharks, pelagic bony fish like the tuna, and dolphins. Second, the scale is not large enough. The fact that eyes have often evolved does not show that had, say, the earliest chordates succumbed to a bit of bad luck (and become extinct), then vertebrate-like organisms would have evolved again. Moreover, Gould's main concern is not with adaptive complexes (the source of Conway Morris' examples) but with body plans – basic ways of assembling organisms. I think we have to score Gould's contingency claims as: 'Don't know; and at this stage don't know how to find out.'

Gould seems clearly right to think that mass extinctions have played a fundamental role in shaping the tree of life, and is probably right in thinking that extinctions work by different rules in mass extinction regimes. Some of these events were clearly intense enough and sudden enough to transform the biological organisation in the world, and their effects were so fast that the species under the gun had no time to adapt themselves to new conditions. Survival or failure depended on their characteristics at the time at which disaster struck. But it is not clear that mass extinction regimes filter species by virtue of features of the species themselves, rather than properties of the organisms that compose the species. These ideas are difficult to assess, and it is also hard to tell just how fundamentally he and Dawkins disagree. For while Dawkins rejects species selection as an explanation of traits of individual organisms (it is too slow), for all he cares, species selection might explain why one lineage is more disparate or species rich than another. My bet is that Gould may well be right in thinking that survival and extinction in mass extinction depends on properties of the species themselves, not just properties of the organisms that compose the species. However, it has proved hard to find really clear, empirically well-founded examples to back up this hunch.

At one stage, sexual reproduction was thought to be maintained by species selection. Sex has a huge cost at the individual level; a cost gene selection makes very vivid. From the perspective of every other gene in the genome, the gene coding for sexual reproduction, the sex gene, is a horrendous outlaw. For it *halves* their chances of replication in any given act of reproduction. Organisms that reproduce asexually copy all of their genes to each offspring; sexual reproducers only half. But perhaps asexual species are selected against. For as species they lack the evolutionary potential of sexual species. With the exception of an obscure group of invertebrates (the *bdelloid rotifers*), species of asexual animals do not seem to last for long. So perhaps sexual reproduction is preserved by the engaging tendency of asexuals to die out in the face of environmental change. This idea has recently fallen on hard times. New ideas of the individual advantage conferred by sex have been developed. Moreover, the idea has a problem: sex does not always promote evolvability. For sex can break up as well as create advantageous gene combinations. If you are particularly well adapted, and you reproduce sexually, you will probably produce less well-adapted offspring. Sex will break up your particularly good gene combinations. So it has been hard to find really convincing examples of species-level properties that are built by species-level selection. The problem is to find: (i) traits that are aspects of species, not of the organisms making up the species; (ii) traits that are relevant to extinction and survival; and (iii) traits that are transmitted to daughter species, granddaughter species and so forth. As we saw earlier, transmission to daughter species is especially problematic. It is not hard to come up with good candidates for traits that matter to a species' prospects. A species' geographic and ecological range, its population structure and the extent of its genetic variation uncontroversially satisfy (i) and probably (ii). But are they transmitted to daughter and grand-daughter species? In short, the idea that species themselves are selected is plausible, but it awaits clear confirmation.

It's time to put my cards on the table. My own views are much closer to those of Dawkins than they are to Gould's. In particular, I think Dawkins is right about microevolution: about evolutionary change within local populations. But macroevolution is not just microevolution scaled up. Gould's palaeontological perspective offers real insights into mass extinction and its consequences, and, perhaps, the nature of species and speciation. Moreover, I think he is right to expand the explanatory agenda of evolutionary biology. While we do want explanations of adaptive complexity, there are also puzzling large-scale patterns in the history of life. So, Dawkins is right about evolution on local scales, but maybe Gould is right about the relationship between events on a local scale, and those on the vast scale of palaeontological time.

Suggested Reading

Chapter 1

Good general and impartial introductions to evolutionary theory are not thick on the ground. One very simple introduction based on Darwin himself is Ernst Mayr's *One Long Argument* (Penguin, 1991). A more technical but still readable treatment is John Maynard Smith's *The Theory of Evolution* (Cambridge University Press, 1993). Mark Ridley's *Evolution* (Blackwell, 1996) is widely regarded as a classic; a third edition appeared in 2004. But it is demanding, and Ridley is a former student of Dawkins, so his view of the subject is influenced by him. Richard Fortey's *Life: A Natural History of the First Four Billion Years of Life on Earth* (Vintage, 1999) is a very enjoyable narrative of evolutionary history.

Gould's review of Dennett appeared in two parts in the *New York Review of Books* in the issues of 12 June and 26 June (1997); there was a heated exchange of letters in the issue of 9 October. Gould reviewed Dawkins' *Climbing Mount Improbable* in *Evolution* (vol. 51, pp. 1020–4, 1997); Dawkins reviewed Gould's *Full House* in the same issue.

Dawkins develops his views of evolution in *The Extended Phenotype* (Oxford, 1982). This is certainly his best book and it is readable, though more demanding than his other works. His ideas are stated more succinctly in *The Selfish Gene* (Oxford, 1989, second edition) and developed further in *Climbing Mount Improbable* (W.W. Norton, 1996). *The Ancestor's Tale* covers some of the same ground as Fortey's *Life*, but is full of brilliant brief asides on theoretical issues in biology. Gould has developed his views much more through his articles than in his books. These are published both in the technical literature and in his *Natural History* column 'This View of Life'. Most of these have come out in his anthologies, but perhaps the best single presentation of his ideas is his *Wonderful Life: The Burgess Shale and the Nature of History* (W.W. Norton, 1989). His *The Structure of Evolutionary Theory* (Harvard University Press, 2002) is quite clear page by page, but it is so enormous that only the truly dedicated can work through it, and it is very easy to lose yourself among the details. In 2005, *Paleobiology* published a special issue (Supplement to vol. 31 (2)) on Gould and his work: Elisabeth Vrba and Niles Eldredge (eds.), *Macroevolution: Diversity, Disparity, Contingency*, and in 2006, Oxford University Press published a collection on Dawkins: Alan Grafen and Mark Ridley (eds.), *Richard Dawkins: How a Scientist Changed the Way We Think*.

Chapter 2

The evolutionary transition from the first lifelike structures to the first organisms is a hot topic in recent evolutionary theory. John Maynard Smith and Eörs Szathmáry's *The Major Transitions in Evolution* (Freeman, 1995) is a modern classic but tough going for the chemistry-challenged. Their *Origins of Life: From the Birth of Life to the Origins of Language* (Oxford University Press, 1999) is a lot easier going but still somewhat demanding. Perhaps the most accessible introduction to these issues is Mark Ridley's *Mendel's Demon: Gene Justice and the Complexity of Life* (Weidenfeld and Nicholson, 2000), though there is a good brief treatment at the end of *The Ancestor's Tale*. Nick Lane's *Power, Sex, Suicide: Mitochondria and the Meaning of Life* (Oxford University Press, 2005) is an original and impressive account of the evolution of complex life. Lane argues that the evolution of the mitochondria was both highly improbable and the essential foundation of complex life. Leo Buss' *The Evolution of Individuality* (Princeton, 1987) is a superb treatment of the transition from single-celled to multi-celled life. Richard Michod's *Darwinian Dynamics: Evolutionary Transitions in Fitness and Individuality* (Princeton University Press, 1999) is an important contribution to these issues, but it is in places quite technical. Stuart Kauffman has developed a view of the origin of life which is closer to a 'cell-first' model than a 'replicator-first' model. His *The Origins of Order: Self-Organization and Selection in Evolution* (Oxford University Press, 1993) is brutally difficult, but he returns to these themes much more gently in *At Home in the Universe* (Oxford University Press, 1995). William Schopf gives a palaeontologist's view of these issues in his *The Cradle of Life: The Discovery of Earth's Earliest Fossils* (Princeton University Press, 1999); Schopf has also recently edited a collection on the same issue for University of California Press: William Schopf (ed.), *Life's Origin: The Beginnings of Biological Evolution.* Simon Conway Morris' *Life's Solution* argues that we still have no good account of how life could begin. Andrew Knoll's *Life On a Young Planet* (Princeton University Press, 2003) covers some of the same ground, but takes the story up to and including the first evolutionary radiation of the animals, in the Cambrian Period. It is superb.

Chapter 3

Dawkins' *The Blind Watchmaker* (W.W. Norton, 1986) and *Climbing Mount Improbable* (W.W. Norton, 1996) are both superb expositions of cumulative selection and its importance. The model of the evolution of the eye is from Nilson and Pelger, 'A Pessimistic Estimate of the Time Required for the Eye to Evolve', *Proceedings of the Royal Society, B* (vol. 256, pp. 53–8, 1994). In *Evolution in Four Dimensions* (MIT Press, 2005), Eva Jablonka and Marion Lamb accept that

Dawkins is right in thinking the replicator principle is of enormous importance, but argue that there are four evolutionarily important inheritance mechanisms. There is no gentle introduction to the complexities of the relationship between genes and organisms, and the implications of these complexities for Dawkins' picture. I have had a go at explaining these issues in chapter 5 of Kim Sterelny and Paul Griffiths' *Sex and Death: An Introduction to the Philosophy of Biology* (University of Chicago Press, 1999), and there is a further development of debate by John Maynard Smith's 'The Concept of Information in Biology' in the June 2000 issue of *Philosophy of Science*. Elliott Sober presents his sceptical response to gene selection in his *The Nature of Selection* (MIT Press, 1984) and *Philosophy of Biology* (Westview, 1993). Gould develops a similar, though simpler, critique in 'Caring Groups and Selfish Genes', one of the papers in *The Panda's Thumb* (Penguin, 1980), and in *The Structure of Evolutionary Theory*, especially pp. 632–8. Joram Piatigorsky's *Gene Sharing and Evolution: The Diversity of Protein Function* (Harvard University Press, 2007) reveals the full complexity of the relationship between genes, gene expression and protein. Matt Ridley's *Genome* (HarperCollins, 2000) is an enjoyable introduction to some of the complexities of the way genes help build bodies; in this case, human bodies. His *Nature Via Nurture* (Fourth Estate, 2003) takes up similar issues about human personalities. (Matt Ridley is not to be confused with Mark Ridley, a student of Dawkins who also writes on evolution.) Pat Bateson and Paul Martin in *Design for a Life: How Behavior and Personality Develop* (Jonathon Cape, 1999) explore some similar issues, but from a very different perspective, as does Gary Marcus in his *The Birth of the Mind* (Basic Books, 2004).

Chapter 4

Real magpies are even more complex than my discussion suggests. The full story is told in *Darryl Jones Magpie Alert* (University of NSW Press, 2002). Austin Burt and Robert Trivers' *Genes in Conflict: The Biology of Selfish Genetic Elements* (Harvard University Press, 2005) is a detailed review of outlaw genes in all their various forms. It is fairly technical, but I know of no good but completely non-technical introduction to outlaw genes. But two good recent reviews are Werren and Beukeboom, 'Sex Determination, Sex Ratios and Genetic Conflict', *Annual Review of Ecology and Systematics* (vol. 29, pp. 233–61, 1998) and G.D. Hurst and J.H. Werren, 'The Role of Selfish Genetic Elements in Eukaryotic Evolution', *Nature Reviews Genetics* (vol. 2, pp. 597–606, August 2001). Haig gives an overview of his views on conflict within the genome in 'Genomic Imprinting and Kinship: How Good is the Evidence?', in *Annual Review of Genetics* (vol. 18, pp. 553–85, 2004). On the role of mitochondria in turning off maleness in plants, see Saumitou-

Laprade and Cuguen, 'Cytoplasmic Male Sterility in Plants', *Trends in Ecology and Evolution* (vol. 9, pp. 431–5, 1994). Armand Leroi, Vassiliki Koufopanou and Austin Burt describe a ghoulish example of a selfish cell lineage in dogs, a cancer that has escaped the death of the body in which it first evolved to become an infective sexual pathogen, in 'Cancer Selection', *Nature Reviews* (vol. 3, pp. 226–30, March 2003). These papers are written for professional audiences but they are reasonably accessible.

Dawkins himself has written the best work on extended phenotypic effects in genes, in *The Extended Phenotype*. But the three examples I give can be followed up in Gould, 'The Triumph of the Root-heads', *Natural History* (vol. 105, pp. 10–17, 1995); Sober and Wilson, *Unto Others: The Evolution and Psychology of Unselfish Behaviour* (Harvard University Press, 1998) and Werren, 'Genetic Invasion of the Insect Body Snatchers', *Natural History* (vol. 103, pp. 36–8, 1994), or, for a more recent but more technical treatment, Bourtzis and O'Neill, 'Wolbachia Infections and Arthropod Reproduction', *Bioscience* (vol. 48, pp. 287–93, 1998). Claude Combes' *The Art of Being a Parasite* (University of Chicago Press, 2005) is not written from an extended phenotype perspective, but it is a brilliant overview of parasite lifestyles and their evolution. Scott Turner's *The Extended Organism* (Harvard University Press, 2000) is a splendid, and splendidly clear, application of the extended phenotype theory to animal-built structures.

Chapter 5

Lee Dugatkin's *Cooperation Among Animals: An Evolutionary Perspective* (Oxford University Press, 1997) is a good survey of the theory of animal cooperation, tied to plenty of actual examples. His *Cheating Monkeys and Citizen Bees: The Nature of Cooperation in Animals and Humans* (Free Press, 1999) covers similar material much less technically. His *The Altruism Equation* (Princeton University Press, 2006) is a historical treatment of these same issues. G.C. Williams' *Adaptation and Natural Selection* (Princeton University Press, 1966) is the classic critique of group selective explanations of cooperation. Sober and Wilson's *Unto Others* (Harvard University Press, 1998) is an important attempt to revive those explanations. John Maynard Smith is responsible for applying game theory to the evolution of social behaviour. He develops this theory in *Evolution and the Theory of Games* (Cambridge University Press, 1982), but this is a technical work. Also technical, but much less demanding are Brian Skyrms' books linking game theory to the evolution of human cooperation: *Evolution of the Social Contract* (Cambridge University Press, 1996) and *The Stag Hunt and the Evolution of Social Structure* (Cambridge University Press, 2003). Peter Hammerstein's edited collection *Genetic and Cultural Evolution of Cooperation* (MIT Press, 2003) is important. It too is written for a professional audience, but most of the papers are reasonably accessible. Robert Axelrod's *The Evolution of Cooperation* (Basic Books, 1984), Sigmund's *Games of Life: Explorations in Ecology, Evolution and Behaviour* and David Barash's *The Survival Game* (Times Books, 2003) are thoroughly readable alternatives. The theory of kin selection was developed by William Hamilton. These papers are in his *Narrow Roads of Gene Land*, vol. 1 (Freeman, 1996), and while the papers themselves are extremely demanding, the central ideas are explained with great clarity in his retrospectives on the papers. Amotz and Avishag Zahivi, in *The Handicap Principle* (Oxford University Press, 1997) develop a different suggestion, arguing that apparently altruistic acts are advertisements of the agent's high genetic quality."

Chapter 6

The interaction between developmental mechanisms, variation and evolution is one of the most active fields in contemporary research. Rudy Raff's *The Shape of Life* (University of Chicago, 1996) is a wonderful introduction. It is intended for a professional audience, but it is beautifully written (and it comes with a glossary). So it is accessible to non-experts. Wallace Arthur's *The Origin of Animal Body Plans* (Cambridge University Press, 2000) is also an important study of these issues. Three excellent recent books written for a non-technical audience on these issues are: Wallace

Arthur, *Biased Embryos and Evolution* (Cambridge University Press, 2004); Marc Kirscher and John Gerhart, *The Plausibility of Life* (Yale University Press, 2005) and Sean Carroll, *Endless Forms Most Beautiful* (W.W. Norton, 2005). For an extreme version of the view that the supply of variation is so highly constrained that there is little for selection to do, see Brian Goodwin's *How the Leopard Changed Its Spots* (Charles Scribner, 1994); Goodwin's views are much more extreme than Gould's on this matter. Bill Wimsatt's work on the relationship between selection, development and evolution is path breaking. Most of it has been published in out of the way places, but he has now collected his most important work into *Re-engineering Philosophy For Limited Beings* (Harvard University Press, 2007).There is no completely non-technical introduction to the methodological issues involved in testing evolutionary hypotheses, though Griffith and I do our best to keep it simple in chapter 10 of our *Sex and Death* (University of Chicago Press, 1999). Orzack and Sober's *Adaptation and Optimality* (Cambridge University Press, 2001) is a recent set of papers on these issues. The shot across the bows that began much of this is Gould and Lewontin's 'The Spandrels of San Marco and the Panglossian Paradigm: A Critique of the Adaptationist Programme', in *Proceedings of the Royal Society of London, B* (vol. 205, pp. 581–98, 1979). It is written with great verve, and is by no means technical. Lisa Lloyd pursues these issues through a great case study, the female orgasm, in her *The Case of the Female Orgasm* (Harvard University Press, 2005), arguing that the female orgasm is probably a developmental side effect of the male orgasm, and is not itself directly the result of selection.

Chapter 7

As Gould sees it, extrapolationism in biology is a continuation of a cluster of methodological ideas from 19th-century geology known as 'uniformitarianism'. Uniformitarian ideas were formulated by Lyall, and taken up by Darwin, and hence have been part of evolutionary biology since its beginning. Gould's *Time's Arrow, Time's Cycle* (Penguin, 1988) is a fine study of uniformitarianism. Two recent and important papers on extrapolationism are: Gould, 'A Task for Paleobiology at The Threshold of Majority', in *Paleobiology* (vol. 21, pp. 1–14, 1995), and 'The Necessity and Difficulty of a Hierarchical Theory of Selection', in Anne Magurran and Robert May (eds.), *Evolution of Biological Diversity* (Oxford University Press, 1999). It is the central preoccupation of his *The Structure of Evolutionary Theory*.

Jonathan Weiner's *The Beak of the Finch* is a terrific exposition of selection and evolution on micro-evolutionary scales. That study documented season-by-season selection on finches scattered across the Galapagos Islands. Dawkins returns to the relationship between macroevolution and microevolution at the very end of *The Ancestor's Tale*, defending some version of the idea that macroevolution reduces to microevolution.

Chapter 8

Eldredge's *Time Frames: The Rethinking of Darwinian Evolution and the Theory of Punctuated Equilibria* (Simon and Schuster, 1985) is a good place to begin on punctuated equilibrium, not least because it reprints his and Gould's original article. He updates his take on the debate in *Reinventing Darwin* (John Wiley, 1995), and again in 'The Sloshing Bucket: How the Physical Realm Controls Evolution' in Crutchfield and Schuster (eds.), *Evolutionary Dynamics: Exploring the Interplay of Selection, Accident, Neutrality, and Function* (Oxford University Press, 2003). Gould reappraises the issues as he sees them in his 'Punctuated Equilibrium Comes of Age', in *Nature* (vol. 366, pp. 223–7, 1993), and at very great length in chapter 9 of *The Structure of Evolutionary Theory* (that one chapter is almost 300 pages). Robert Carroll surveys the empirical evidence for the punctuated equilibrium pattern in his *Patterns and Processes of Vertebrate Evolution* (Cambridge University Press, 1999). John Thompson, in *The Coevolutionary Process* (University of Chicago Press, 1994) documents many examples of evolutionary changes in local populations. Elisabeth

Vrba develops her 'turnover pulse hypothesis' in 'Turnover-Pulses, The Red Queen and Related Topics', in the *American Journal of Science* (vol. 293 A, pp. 418–52, 1993). Ernst Mayr's views on speciation are given most succinctly in the relevant section of his two volumes of essays, *Evolution and the Diversity of Life* (Harvard University Press, 1976) and *Toward a New Philosophy of Biology* (Harvard University Press, 1988). For a very sceptical view of punctuated equilibrium and its significance, see Daniel Dennett's *Darwin's Dangerous Idea* (Simon and Schuster, 1995).

Chapter 9

Derek Briggs and Peter Crowther's *Palaeobiology II* (Blackwell, 2001) is a terrific general resource on palaeobiological issues. Douglas Erwin's *The Great Paleozoic Crisis* (Columbia University Press, 1993) is a splendid overview of the Permian extinction. He updates the story in *Extinction: How Life On Earth Nearly Ended* (Princeton University Press, 2006). Archibald's *Dinosaur Extinction and the End of An Era* (Columbia University Press, 1996) is a judicious discussion of the debate about the most contentious extinction of them all. *Evolutionary Paleobiology*, a collection edited by Jablonski, Erwin and Lipps (University of Chicago Press, 1996), has many papers relevant to these issues. These three books are written for professional audiences, though those by Erwin and Archibald, especially, are clear and well written. Peter Ward has written two enjoyable and non-technical books on mass extinction. They are: *On Methuselah's Trail: Living Fossils and the Great Extinctions* (Freeman, 1992) and (the rather preachy) *The End of Evolution* (Bantam, 1994). Ward is sympathetic to Gould's take on these issues. Michael Benton's *When Life Nearly Died* (Thames and Hudson, 2003) is a more recent look at the Permian extinction, and is written for a non-technical audience. Richard Fortey's *Life: A Natural History of the First Four Billion Years* (Vintage, 1999) is good on these issues, too. David Raup analyses extinction, and particularly whether mass extinction is a fair game, in his *Extinction: Bad Genes or Bad Luck?* (Oxford University Press, 1991). Richard Bambach, Andrew Knoll and Steve Wang argue in 'Origination, Extinction, and Mass Depletions of Marine Diversity', *Paleobiology* (vol. 30, pp. 522–42, 2004) that some of the classic mass extinctions are actual slowdowns of origination.

Gould writes on mass extinction in all of his *Natural History* collections. In these, he mostly emphasises the discontinuities in evolutionary history that mass extinction causes. He explores the idea that mass extinction imposes a filter on species, rather than the individuals that make up the species, in 'A Task for Paleobiology', *Paleobiology* (vol. 21, pp. 1–14, 1995); 'The Necessity and Difficulty of a Hierarchical Theory of Selection' in Magurran and May's *Evolution of*

Biological Diversity (Oxford University Press, 1999); and in two papers written jointly with Lisa Lloyd. These are 'Species Selection on Variability' and 'Individuality and Adaptation Across Levels of Selection', both published in *Proceedings of the National Academy of Science* (vol. 90, pp. 595–9, 1993 and vol. 96, pp. 11,904–09, 1999). And, especially, in chapter 8 of *The Structure of Evolutionary Theory*.

Chapter 10

Gould makes his case about the Cambrian and its significance in *Wonderful Life: The Burgess Shale and the Nature of History* (W.W. Norton, 1989), and in it he makes a pre-emptive strike against Dawkins' line of criticism. These issues were followed up in a series of specialist articles in *Paleobiology*: Dan McShea, 'Arguments, Tests, and the Burgess Shale' (vol. 19, pp. 399–402, 1993); Mark Ridley, 'Analysis of the Burgess Shale' (vol. 19, pp. 519–21, 1993) to which Gould replies, and especially Gould's 'The Disparity of the Burgess Shale Arthropod Fauna: Why We Must Strive to Quantify Morphospace' (vol. 17, pp. 411–23, 1991). In these papers, Gould explores an additional way of understanding the contrast between Cambrian and post-Cambrian animal life; the distinction between a fauna with a relatively open and flexible developmental system, and a fauna with a less open, more rigid system. Mark and Dianna McMenamin argue that selection was important in generating disparity in *The Emergence of Animals* (Columbia University Press, 1991). They suggest that the Cambrian explosion is a response to the invention of predation. Andrew Parker, in *In the Blink of An Eye* (Basic Books, 2003), suggests an alternative ecological hypothesis. James Valentine's *On the Origin of Phyla* (University of Chicago Press, 2005) is a synthesis of what we know about early animal evolution.

Simon Conway Morris, himself one of those who reinterpreted the Burgess Shale, takes issue with Gould in *The Crucible of Creation* (Oxford University Press, 1998). He returns to these issues in *Life's Solution: Inevitable Humans in a Lonely Universe* (Cambridge University Press, 2003). In both books, he argues that the history of life is much less contingent than Gould supposes, and that Gould overstates the weirdness of the Burgess fauna. Graham Budd and Soren Jensen use the distinction between crown group members of a lineage and stem group members of that lineage to argue that Cambrian disparity is overestimated in their 'A Critical Reappraisal of the Fossil Record of the Bilaterian Phyla' (*Biological Reviews of the Cambridge Philosophical Society*, vol. 75, pp. 253–95, 2000). Guy Narbone splendidly reviews what we know about the Ediacarans in his 'The Ediacara Biota: Neoproterozoic Origin of Animals and Their Ecosystems' (*Annual Review of Earth Sciences*, vol. 33, pp. 421–42, 2005). Charles Marshall's 'Explaining the Cambrian Explosion of Animals', *Annual Review of Earth Sciences* (vol. 34, pp. 355–84, 2006) is an intelligent, clear and fair review of the main hypotheses about that great radiation of animal life. Lindell Bromham's 'Molecular Dates for the Cambrian Explosion: Is the Light at the End of the Tunnel an Oncoming Train?' is a good and brief overview of molecular clocks and their reliability. It is published electronically, in *Palaeontologia Electronica* (vol. 9, 2006).

For a good recent introduction to cladistics – the view that leads to Ridley's scepticism about the diversity/disparity distinction – see Henry Gee's *In Search of Deep Time: Beyond the Fossil Record to a New History of Life* (Free Press, 1999). Joel Cracraft and Michael Donoghue (eds.), in *Assembling the Tree of Life* (Oxford University Press, 2004), give a good overview of current opinion about the family tree of life, though how long it will stay current is anyone's guess.

Chapter 11

The key text from Gould on these issues is *Full House* (Harmony Books, 1996), though these themes have been explored in his *Natural History* essays for years, often using baseball as a model system for discussing the importance of changes in variation in a system. John Maynard Smith and Eörs Szathmáry discuss evolutionary transitions in both *The Major Transitions in Evolution* (Oxford University Press, 1995) and *Origins of Life* (Oxford University Press, 1999). Dawkins discusses progress in his review of *Full House*, in 'Human Chauvinism', *Evolution* (vol. 51, pp. 1015–20, 1997), in 'Progress', an essay in Evelyn Fox Keller and Elisabeth Lloyd (eds.), *Key Words in Evolutionary Biology* (Harvard University Press, 1992), and at the end of *The Ancestor's Tale*. Daniel Dennett defends the idea that life increases in adaptiveness over time in *Darwin's Dangerous Idea* (Simon and Schuster, 1995). J.T. Bonner defends the idea that there is a real, selection-driven trend for an increase in complexity over time in *The Evolution of Complexity by Means of Natural Selection* (Princeton University Press, 1988). Geerat Vermeij develops an alternative version of this idea in his *Nature: An Economic History* (Princeton University Press, 2004). Michael Ruse puts all these debates into their historical context in *Monad to Man: The Concept of Progress in Evolutionary Biology* (Harvard University Press, 1996).

Chapter 12

Gould's views on the relationship between science and religion are explored in his *Rocks of Ages* (Ballentine, 1999); Dawkins tells his very different story in *The God Delusion* (Houghton Mifflin, 2006). Dawkins argues for the idea that scientific knowledge is liberating in his *Unweaving the Rainbow* (Penguin, 1998). His views on memes are most developed in the essays in *A Devil's Chaplain* (Houghton Mifflin, 2003). The hunch that human evolution depends as much on memes as on genes is explored most systematically not by Dawkins but by Dennett, in *Darwin's Dangerous Idea* (Simon and Shuster, 1995); he applies them to religion in his *Breaking the Spell: Religion as a Natural Phenomenon* (Allen Lane, 2006). Peter Richerson and Robert Boyd's *Not By Genes Alone: How Culture Transformed Human Evolution* (University of Chicago Press, 2005) is a compelling introduction to the more general project of applying evolutionary models to human culture. The best response to meme theory is Dan Sperber's *Explaining Culture* (Blackwell, 1996). For a provocative, but in my view seriously mistaken, application of evolutionary thought about humans, see Thornhill and Palmer's *A Natural History of Rape* (MIT Press, 2000). There is much better work available than this, especially that which locates human social evolution in its great-ape context. Two good recent examples are Michael Tomasello's *The Cultural Origins of Human Cognition* (Harvard University Press, 1999) and Chris Boehm's *Hierarchy in the Forest: The Evolution of Egalitarian Behavior* (Harvard University Press, 1999). Sarah Blaffer Hrdy's *Mother Nature: Natural Selection and the Female of the Species* (Pantheon Books, 1999) is also in places very speculative. But its account of the action of selection on human sexuality is much more subtle than that of Thornhill and Palmer. So too is Garth Fletcher's *The New Science of Intimate Relationships* (Blackwell, 2002), which takes seriously both variation within the sexes and similarities between them. I have had my own go at human cognitive evolution in *Thought in a Hostile World* (Blackwell, 2003). Kevin Laland and Gillian Brown's *Sense and Nonsense: Evolutionary Perspectives on Human Behaviour* (Oxford University Press, 2002) is both clear and fair as it surveys the main lines of thought on these issues. David Buller's *Adapting Minds: Evolutionary Psychology and the Persistent Quest for Human Nature* is a beautifully written critique of one of these main lines, that exemplified by the approach of Thornhill and Palmer.

Glossary

adaptation: an adaptation is a characteristic of an organism that exists today because it helped that organism's ancestors survive or reproduce.

adaptive trait: a trait that helps an organism with that trait survive or reproduce.

allele: an alternative version of a gene. Genes are located at particular regions of a chromosome. In a particular population, there may be different versions of a gene at a given location. These alternative versions are the alleles of that gene at that location.

amino acids: the building blocks of proteins. The genetic code specifies amino acids in a system that relates a sequence of three DNA bases to a single amino acid.

arms race: evolutionary interactions, within a species or between two species, in which each player becomes better adapted as a result of interaction with the other player.

biota: the totality of living things in a region or at a time.

chromosome: a long sequence of genes joined together in DNA molecules built around structurally supporting proteins. Chromosomes occur only in eukaryotic organisms. The number of chromosomes varies across species, but all (normal) members of a given species will have the same number.

clade: a lineage consisting of all of a group of species and their common ancestor.

diploid cell: a cell that has two versions of each chromosome. If the organism is the result of sexual reproduction, each parent provides one of each pair of chromosomes.

ethology: the evolutionary study of animal behaviour in the wild, rather than its study under unusual, laboratory, conditions.

eukaryotes: organisms built from complex eukaryotic cells. Each cell has a discrete nucleus, together with complex cellular machinery usually including mitochondria and, in plants, chloroplasts. Eukaryotic cells are thought to have arisen from the evolutionary fusion of bacterium-like organisms. Mitochondria and chloroplasts had free-living bacteria as ancestors.

fitness: a measure of the probability that an organism (or a gene, or a group) will reproduce itself. Comparative fitness is of particular evolutionary significance: the evolutionary history of a population will depend on which organisms (or genes, or groups) do better than others.

gamete: the sex cell of an organism (e.g. sperm, ova,

pollen). It is haploid, having half the chromosome number typical of the species, and fuses in sexual reproduction with another gamete to restore the full set for the species.

gene: a DNA sequence. The exact definition of a gene remains a matter of controversy, but genes are DNA sequences of some kind. The debate is whether each gene must have an identifiable function, or whether the DNA sequences can be of arbitrary length, and with arbitrary boundaries.

genome: the total collection of genes that an organism carries.

genotype: often used as a synonym for 'genome'. But it is sometimes used to specify the genes an organism has at a specific region (or regions) of a chromosome.

haploid cell: a cell that has only a single set of chromosomes.

heritability: a measure of the probability that an offspring will share a trait possessed by its parent (at least in mathematical versions of evolutionary theory). A trait is heritable if a parent's having that trait increases the probability that its offspring will also have it.

macroevolution: a series of evolutionary changes in one or more species lineages; typically large, very long-lasting species lineages.

meiosis: the special form of cell division which generates sex cells, each of which has only half the number of chromosomes typical of cells of that species. This is in contrast to standard (asexual) cell division, whereby daughter cells end up with copies of all the structures in the parent cell.

microevolution: evolutionary changes within a single species. The term is sometimes used to refer to the evolution of one species into its immediate descendant(s).

mitochondrion: a special structure in eukaryote cells that generates energy for the cell, and has its own DNA. This DNA is almost always inherited only through the female line.

monophyletic group: a group that contains: (i) an ancestor species, (ii) only the descendants of that ancestor and (iii) all the descendants of that ancestor.

mutation: a new DNA sequence that is produced when an error occurs in the copying process of a gene (or another replicator), resulting in a difference between the daughter gene and the template from which it was copied. Mutations are one source of new genetic variation in the population. Most, if they have any effects, have bad ones. So selection has acted to make the copying process very accurate indeed. But organisms have so many genes that even accurate copying still generates appreciable numbers of mutations.

natural selection: the process by which the superior fitness of certain traits causes those traits to increase in frequency in a population.

phenotype: an organism's developed morphology, physiology and behaviour. It contrasts with the genotype: the genes that an organism carries.

prokaryotes: single-celled organisms, such as bacteria, without a nucleus or mitochondria. Prokaryotes are the simplest and oldest forms of life. There are two great divisions of prokaryotes: bacteria and archaea. The division between these lineages is the oldest split in the tree of life.

protein: a very large molecule made up of chains of amino acids folded in extraordinarily complex ways.

replicator: a structure that causes copies of itself to be made, and that, in combination with others, sometimes constructs a vehicle of selection. It is Dawkins' unit of heredity and selection.

species: there is no uncontroversial definition of a species. The most usual definition is the 'biological species concept', which defines a species as an inter-breeding population of organisms. But there are many problems in making this notion precise. More-over, using this definition, no asexual organisms form species.

species sorting: any pattern in species survival or extinction counts as species sorting, whatever the cause of that pattern. If, for example, for whatever reason, species with small population sizes are at a greater risk in mass extinction events, that would count as species sorting.

vehicle: a structure built by gene combinations in

development. A vehicle mediates the reproduction of the genes responsible for its production. The clearest examples of vehicles are individual organisms, but there may be others, including groups of organisms.

Appendix: Geological Timescale

Era	Period	Epoch	Duration
Cenozoic	Quaternary	Holocene	12,000 BP to present
		Pleistocene	1.8 mya–12,000 BP
	Tertiary	Pliocene	5–1.8 mya
		Miocene	23–5 mya
		Oligocene	34–23 mya
		Eocene	56–34 mya
		Palaeocene	65–56 mya
Mesozoic	Cretaceous		145–65 mya
	Jurassic		200–145 mya
	Triassic		251–200 mya
Palaeozoic	Permian		299–251 mya
	Carboniferous		359–299 mya
	Devonian		416–359 mya
	Silurian		444–416 mya
	Ordovician		488–444 mya
	Cambrian		542–488 mya
Precambrian	*Various*		*lower limit not defined*–542 mya

BP = years before present
mya = millions of years ago
All figures are approximate

The stretch from the Cambrian to now is known as the Phanerozoic Eon. The Precambrian is divided into Eons too, though their limits are not very precisely defined. The Proterozoic Eon lasted from about 2.5 billion years ago to the base of the Cambrian at about 542 million years ago. It was preceded by the Archaen Eon, stretching back to about 3.8 billion years ago, which in turn was preceded by the Hadean Eon, stretching back to the formation of the Earth, over 4 billion years ago."

The Comet Sweeper

Caroline Herschel's Astronomical Ambition

Claire Brock

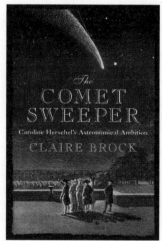

The incredible story of Britain's first female professional scientist – a true Enlightenment celebrity long overdue for a rediscovery.

Such was Caroline Herschel's reputation that a congratulatory letter in 1790 from the director of the Paris Observatory was simply addressed to 'Mlle Caroline Herschel, Astronome Célèbre, Slough.'

Having escaped domestic servitude in Germany by teaching herself to sing and establishing a career in England, Herschel learned astronomy while helping her brother William, then royal astronomer. Soon she was making scientific discoveries in her own right, and she swept to international scientific and popular fame. She was awarded a salary by George III in 1787, becoming the first woman in Britain ever to make her living from science.

But, as a woman in a male-dominated world, Herschel's great success was achieved despite constant frustration of her ambitions. Assisting her brother had to take priority over her own work, and his marriage separated her from the instruments of her trade, stalling her career.

Drawing on original sources – including Herschel's diaries and her fiery letters – Claire Brock tells the story of a woman so determined to win independence and satisfy her ambition that she moved careers and countries while chasing success.

Hardback £9.99

ISBN 10: 1-84046-720-7   ISBN 13: 978-1840467-20-8

Dissent Over Descent

Evolution's 500-Year War on Intelligent Design

Steve Fuller

Iconoclastic and provocative
as ever, Steve Fuller makes
us think again about Intelligent
Design and Evolution.

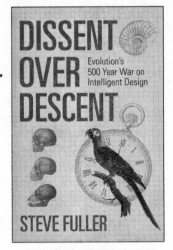

If you think Intelligent Design Theory (IDT) is merely the
respectable face of Christian fundamentalism, and Evolution
the only sensible scientific world-view, think again.
Historically, the situation is almost exactly reversed.

IDT has driven science for 500 years. It was responsible for
the 17th century's Scientific Revolution and has helped build
modern histories of physics, mathematics, genetics and
social science. IDT's proponents have upheld, often
secretly, radical views of Christianity, taking literally the
Biblical idea that humans have been created in God's image.
This confident, even arrogant, view of humanity enabled the
West to triumph in the modern era.

Steve Fuller's brilliant revisionist history is essential reading
for anyone who wants a deeper understanding of science's
most vociferous debate.

Hardback £9.99

ISBN 10: 1-84046-804-1   ISBN 13: 978-1840468-04-5

The Science of Doctor Who

Paul Parsons

Have you ever wondered how Daleks climb stairs? Or where the toilets are on the Tardis? This hugely acclaimed guide tells all …

Since 1963, the journeys of the Time Lord have shown us alien worlds, strange life forms, futuristic technology and mind-bending cosmic phenomena. Viewers cowered terrified of Daleks, were amazed with time travel, and travelled through black holes into other universes and new dimensions.

The breadth and imagination of *Doctor Who* have made the show a monumental science fiction success. *BBC Focus* editor Paul Parsons explains the scientific reality.

Discover:

- why time travel isn't ruled out by the laws of physics
- the real K-9 – the robot assistant for space travellers built by NASA
- how genetic engineering is being used to breed Dalek-like designer life forms
- why before long we could all be regenerating like a Time Lord
- the medical truth about the Doctor's two hearts, and the real creature with five.

Paperback £8.99

ISBN 10: 1-84046-791-6   ISBN 13: 978-1840467-91-8

Why Aren't They Here?

The Question of Life on Other Worlds

Surendra Verma

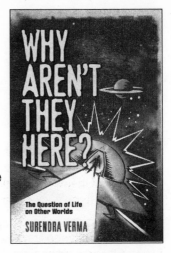

Is there anybody out there? Are there other life-forms lurking in outer space – or are they already here? Surendra Verma investigates …

The rate of expansion of our universe is mind-blowing: imagine a pea growing to the size of the Milky Way in less time than it takes to blink. In all this infinite space that we cannot even see, let alone explore, it seems certain that there is some life on other worlds.

Sir Arthur C. Clarke declared that 'the universe is full of intelligent life – it's just been too intelligent to come here'. Journalist Surendra Verma brilliantly outlines the historical, fictional, speculative and emerging scientific opinions on what alien life might be like.

From Aristotle to ET via radio, religion and reincarnation, this fast-moving narrative examines history and dispels myths before focusing on the possibilities lurking in space. In a popular and easy-to-read style, Verma uses current research to speculate on what life is like on other planets, how we might communicate with them, and what Earth might seem like to visitors.

Hardback £9.99

ISBN 10: 1-84046-806-8   ISBN 13: 978-1840468-06-9

Atom

Piers Bizony

The official tie-in to a brilliant
new BBC TV series, Atom is
the history of the people behind
the greatest human scientific
discovery ever …

No one ever expected the atom
to be as bizarre, as capricious
and as weird as it turned out to
be. Its story is one riddled with jealousy, rivalry, missed
opportunities and moments of genius.

John Dalton gave us the first picture of the atom in the early
1800s. Almost 100 years later came one of the most
important experiments in scientific history, by the young
misfit New Zealander, Ernest Rutherford. He showed that
the atom consisted mostly of space, and in doing so turned
200 years of classical physics on its head.

It was a brilliant Dane, Niels Bohr, who made the next great
leap – into the incredible world of quantum theory. Yet he
and a handful of other revolutionary young scientific Turks
weren't prepared for the shocks Nature had up her sleeve.
Mind-bending discoveries about the atom were destined to
upset everything we thought we knew about reality. Even
today, as we peer deeper and deeper into the atom, it throws
back as many questions at us as answers.

Hardback £12.99

ISBN 10: 1-84046-800-9   ISBN 13: 978-1840468-00-7

The Chilling Stars

A New Theory of Climate Change

Henrik Svensmark and Nigel Calder

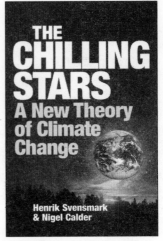

Scientists agree that over the last century the Earth has become warmer. But do we really know why this has happened?

A deftly written and enjoyable read, The Chilling Stars outlines a brilliant, daring and undoubtedly controversial new theory that will provoke fresh thinking about global warming. As prize-winning science writer Nigel Calder and climate physicist Henrik Svensmark explain, an interplay of the clouds, the Sun and cosmic rays – sub-atomic particles from exploded stars – seems to have more effect on the climate than man-made carbon dioxide.

This conclusion stems from Svensmark's research at the Danish National Space Center which has recently shown that cosmic rays play an unsuspected role in making our everyday clouds. And during the last 100 years cosmic rays became scarcer because unusually vigorous action by the Sun batted many of them away. Fewer cosmic rays meant fewer clouds and a warmer world.

The theory, simply put here but explained in fascinating detail in the book, emerges at a time of intense public and political debate about climate change. Motivated only by their concern that science must be trustworthy, Svensmark and Calder invite their readers to put aside their preconceptions about man-made global warming and look afresh at the role of Nature in this hottest of world issues.

Paperback £9.99

ISBN10: 1-84046-815-7   ISBN13: 978-1840468-15-1